The Primate Family Tree

The Amazing Diversity of Our Closest Relatives

Ian Redmond

Foreword by Jane Goodall

Published by Firefly Books Ltd. 2011

First printing

Publisher Cataloging-in-Publication Data (U.S.)

Redmond, Ian.
The primate family tree : the amazing diversity of our closest relatives / Ian Redmond.
Originally published 2008.
[ ] p. : col. photos., maps ; cm.
Includes bibliographical references and index.
Summary: Examines the nature of primates in each branch of the primate family tree, exploring their origins, evolutionary links and differences between different primate groups, primate behavior, social structures, and relationships with humans, primate habitats, their prospects for survival, and conservation issues.
ISBN-13: 978-1-55407-964-3 (pbk.)
1. Primates. I. Title.
599.8 dc22 QL737.P9.R436 2011

Library and Archives Canada Cataloguing in Publication

Redmond, Ian
The primate family tree : the amazing diversity of our closest relatives / Ian Redmond.
Includes bibliographical references and index.
ISBN 978-1-55407-964-3
1. Primates. 2. Primates--Pictorial works. I. Title.
QL737.P9R44 2011 599.8 C2011-901170-0

Published in the United States by
Firefly Books (U.S.) Inc.
P.O. Box 1338, Ellicott Station
Buffalo, New York 14205

Published in Canada by
Firefly Books Ltd.
66 Leek Crescent
Richmond Hill, Ontario L4B 1H1

Conceived, edited and designed
in the United Kingdom by
Marshall Editions
The Old Brewery
6 Blundell Street
London N7 9BH
www.marshalleditions.com

For Marshall Editions
Publisher: Richard Green
Art Director: Ivo Marloh
Commissioning Editor: Claudia Martin
Managing Editor: Paul Docherty
Layout & Editorial: Schermuly Design Co.
Copy Editor: Ben Hoare
Researcher: Kara Moses
Picture Manager: Veneta Bullen
Indexer: Sue Butterworth
Production: Nikki Ingram
Cover design: Ivo Marloh

Originated in Hong Kong by Modern Age
Printed and bound in China by SNP Leefung Printers Limited

Previous page:
A group of Northern Plains Gray Langurs lies basking in the sun on a cliff face in the Thar desert, Rajasthan, India.

Facing page:
A Rhesus Macaque female carefully grooms her infant in Keoladeo National Park, Bharatpur, India.

Following pages:
In Joshin-etsu National Park, Japan, a group of Japanese Macaques stay warm in a hot geothermal spring pool.

CONTENTS

AUTHOR'S NOTE

SPECIES LISTS

Primate taxonomy has seen several dramatic revisions in recent years, with the number of accepted species doubling since the 1970s, and species new to science still being discovered in remote parts of the world every year. Unfortunately, in a book of this size, we have been unable to list every one of the newly described species in the larger groups. In such cases, we have listed as many species as space allowed, and the interested reader is urged to find further information using the online resources and books listed on page 171. Species names have been capitalized in the book, in order to distinguish them from names that refer to groups of species.

CONSERVATION STATUS

The species tables contain information about the conservation status of each primate species, as assessed on the Red List of the World Conservation Union (IUCN). The tables also supply the CITES listing for each species. CITES (the UN Convention on International Trade in Endangered Species of Wild Fauna and Flora) is an international agreement between 172 governments. Its aim is to ensure that international trade in wild animals and plants does not threaten their survival. It lists species in three appendices, according to the degree of protection they need. Appendix I includes species threatened with extinction, and trade in these species is permitted only in exceptional circumstances—more than 100 primate species are listed here. Appendix II includes species not necessarily threatened with extinction, but in which trade must be controlled in order to ensure the survival of the species; all primate species not listed in Appendix I are listed in Appendix II. For details on where to obtain more information about the Red List and CITES, see Further Resources, page 171.

WHERE TO WATCH PRIMATES

If you are reading this, you should seriously consider making your next holiday a primate-watching trip. Every country with primates is a potential destination, but some have yet to develop facilities or offer trained guides. In the species tables, those countries where there is a good chance of seeing a species are marked with a star (✪). Where many species are found in only one country—lemurs, for example—the star denotes those most likely to be seen by an interested visitor. Some of the world's top primate-watching sites are also listed on page 170. Good luck, and may your primate experiences bring you as much joy as they have brought me.

Ian Redmond

ABBREVIATIONS

In the tables that supply the range-states of species, the following abbreviations for countries have been used:

CAR Central African Republic

Congo Republic of Congo, which lies north of the River Congo

DRC Democratic Republic of Congo (formerly Zaire, formerly Belgian Congo), which is bisected by the River Congo

Eq. Guinea Equatorial Guinea (including Rio Muni and Bioko Island)

Foreword

At first sight, a 4.2-ounce Pygmy Marmoset, a 550-pound gorilla, a bushbaby and a human seem to have little in common—yet all four are primates. We are a motley clan indeed, we fellow primates, including in our ranks the huge, red-haired orangutan of Sumatra and Borneo, the Japanese Macaque, which roams snow-covered mountains in Japan, the Hamadryas Baboon of the Ethiopian deserts, the Red Uakari of the Amazon jungle, the Ring-tailed Lemur of Madagascar—and so many more. Humans are most closely related to the great apes, especially the Chimpanzee and Bonobo. We are, in fact, the seventh great ape, although we seldom think of ourselves in this way.

CHANGING ATTITUDES

When, in 1960, I began observing the Chimpanzees of the Gombe Stream Game Reserve (today a national park) on the eastern shores of Lake Tanganyika, there had been very few detailed studies of primates in the wild. My descriptions of their complex social behavior, of tool-making and tool-use that required intelligence and of emotions that seemed similar to our own were largely dismissed as anthropomorphic by western science. Even the use of names rather than numbers to identify the Chimpanzees being studied was frowned upon by many European ethologists of the time. Gradually, though, this rigid attitude began to change, as field studies of animals with complex brains increasingly provided evidence of similarly complex behavior. Now, nearly half a century later, we acknowledge that many behaviors—intellectual, social and emotional—that were once believed to be unique to ourselves are shown by other animals too. Thousands of studies of primates, including humans, have revealed fascinating information about the extent of our relatedness and the evolutionary pathway that has led to the diverse array of primates on our planet today—and this book explores every branch of this family tree of ours.

THE ENDANGERED PRIMATE WORLD

Tragically, even as our knowledge of and respect for other primates has grown, so their numbers have shrunk. While reading this book, you will become aware time and again of how much the future of so many of our primate relatives is threatened by human actions. I hope that its beautifully produced pages will not only fascinate readers but also help them to realize just how endangered so many primates and their habitats are today. They need our help if they are to survive beyond this century; to this end a list of organizations working for primate conservation is included at the end of the book—I encourage you to get involved in any way that you can.

For many of us, the problems of species living far away seem to have little bearing on our lives. Yet the more I travel, raising awareness of the plight of Chimpanzees and other primates, the more I realize the extent to which so many seemingly disparate issues are interconnected. One obvious example is the destruction of the world's rainforests, which are disappearing at a terrifying rate. Yet even though only about 50 percent of the world's original forests remain, forest ecosystems store more than double the total amount of carbon in the atmosphere today. Thus, unless we can halt the destruction of rainforests, our own future—as well as that of the primates and all the other species in the forest—will be ever more acutely in jeopardy as a result of climate change.

MAKING A DIFFERENCE

Many people feel helpless in the face of global problems, not realizing that they can help—yet every one of us can make a difference, however small. As consumers and voters we have enough power to affect business practices and even government regulations. For example, palm oil is found in 10 percent of supermarket products, ranging from everyday foods to cosmetics (on which it is often listed simply as "vegetable oil") and is now also in demand for biofuels. Most of the world's palm oil comes from Malaysia and Indonesia, where plantations have now replaced huge areas of natural forest, once home to orangutans and countless other species. Conservationists, realizing the scale of the crisis, stepped up their efforts to raise awareness. As a result, more and more people began to boycott products containing palm oil. The industry was thus pressured to take palm oil only from established plantations, and products containing palm oil that is clearly identified as "sustainably produced" will become increasingly available. Of course, we can support conservation efforts in other ways, such as writing to our elected representatives or to the companies themselves; and we can always make a donation to those working on the ground.

THE BENEFITS OF ECOTOURISM

Those who want to become involved at a more hands-on level may wish to visit primates in their natural habitats. Even a short encounter with a Chimpanzee, a Golden Lion Tamarin, a howler monkey or

Above Hooting in harmony—an orphan Chimpanzee seems to recognize a kindred spirit in Dame Jane Goodall.

any of the other wonderful primates that you will find in this book can be a life-changing experience. Responsible tourism that respects the wildlife and the culture of the local community may help to focus attention on a particular habitat and the animals living there. It may also provide jobs and a better standard of living for local people.

The most important thing is that we, as citizens of Planet Earth, should do something to help—the cumulative effect of small individual actions can bring about big changes. This is the most important message that I take around the world and the reason I work so hard to develop our Roots & Shoots program for young people, now in nearly 100 countries. We could all wear ourselves out trying to protect primates and their habitats, but it would be of little use if we were not educating today's children to become better stewards of Earth than we have been.

When I was a child I loved books. I used to spend hours going through old volumes on natural history, delighting in the illustrations—black-and-white engravings for the most part. I was especially fascinated by the accounts of monkeys and apes, partly, no doubt, from reading *The Jungle Book* and the Tarzan stories. A book such as *The Primate Family Tree* gives me the same kind of pleasure today. You are in for a real treat.

Jane Goodall PhD, DBE
Founder—Jane Goodall Institute
UN Messenger of Peace
www.janegoodall.org

Jane Goodall

What Is a Primate?

Main image Like other primates, these Vervet Monkeys give maternal protection to helpless infants and use social grooming to cement caring relationships between family and friends.

Facing page Although they have long, doglike muzzles, baboons have close-set, forward-facing eyes, one of several characteristics that set primates apart.

Primate Characteristics

From mouse lemurs to gorillas, the Primates are an extraordinarily diverse and successful Order of mammals. Lemurs, monkeys and apes are the best-known primates, but the Order also includes the tiny tarsiers, which look like nocturnal gremlins, Asian lorises—which resemble animated soft toys but are, in fact, the only poisonous primate—fluffy-tailed African bushbabies, the spiny-necked Potto and the amazing angwantibo, which is able to sniff out a caterpillar from a distance of about 3 ft. (1 m).

There is no single feature that makes an animal a primate, but rather a suite of features that together make primates unmistakable. Size is certainly not a common feature. The smallest species weigh in at under 2 oz. (55 g)—the same as an average tomato or egg. The heaviest primates would tip the scales at 5,000 times that—a silverback gorilla can weigh up to 550 lb. (250 kg). But to list male gorillas as the heaviest primate, as so many books do, is to ignore the diversity of human size range. Humans are primates too, and some sumo wrestlers weigh more than a silverback; in fact, the heaviest individual primates on record are heavyweight humans in excess of 1,100 lb. (500 kg).

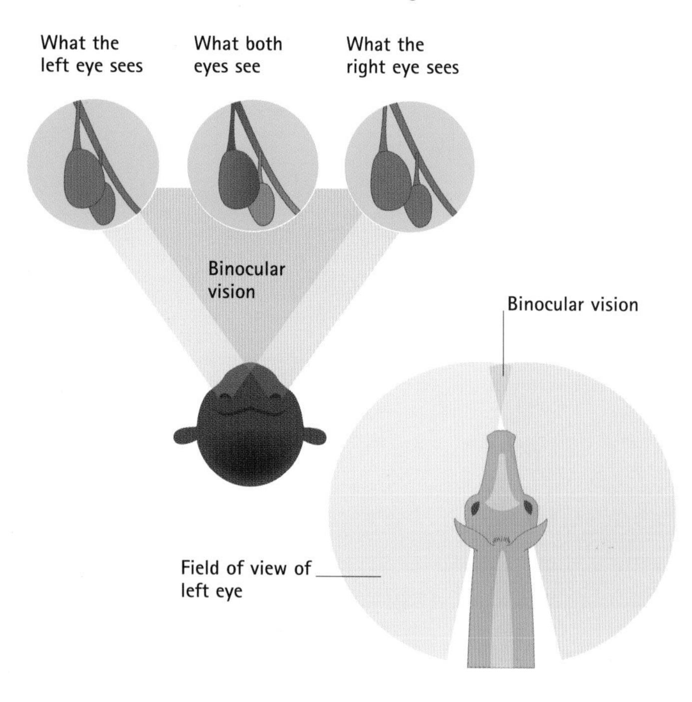

Above Primate eyes face forward and perceive scenes from two slightly different angles. The overlap between these angles produces a three-dimensional image. Animals with sideways-facing eyes, such as horses, have a wider total field of view but less overlap between the scenes viewed—and so less binocular vision.

Above Nocturnal primates, such as this Philippine Tarsier, have large eyes. In bright light, their pupils contract to form tiny slits. In the dark, the pupils expand greatly and the eyes function as huge, black, light-gathering orbs.

SIX KEY CHARACTERISTICS

Most primates possess the following characteristics, although a few species do not possess all of them.

Forward-facing eyes: both eyes point in the same direction, allowing primates to perceive a three-dimensional view of whatever they look at, although they are unable to look at what is behind them without turning their head.

Eye sockets: the eyeball sits inside a solid, protective ring of bone called the orbit.

Grasping hands: one of the digits grips against the other four, enabling a firm grasp.

Nails: fingers and toes end in a flat nail to protect the sensitive tip.

Fingerprints: the skin on the fingertips and underside of the hand is bare and covered in a pattern of tiny ridges, which are unique to each individual.

Large brains: large cerebral hemispheres (compared with other mammals) give primates higher intelligence, the ability to learn and a complex repertoire of behaviors.

Above This Victorian illustration shows the close resemblance between human and gorilla skeletons. One of the most prominent differences is in the lengths of the limbs. The shorter legs and longer arms of gorillas compared to those of their human relatives are adaptations to walking on all fours and climbing in trees.

Specialized vision

One way to study primate anatomy is to look in a mirror and compare your shape with that of a dog or horse. Notice that your eyes face forward with overlapping fields of vision, giving you a three-dimensional view of the world, unobstructed by the muzzle (which is reduced in most primates). This enables you to judge distances accurately—an important ability for a life in the trees, where primates evolved and the ability to jump or swing from branch to branch is paramount. Although primates have lost the all-round vision that sideways-facing eyes deliver, your mobile neck enables you to turn your head to scan your surroundings. Data from all the senses is then processed by a brain that is larger in proportion to body size than in most other mammal groups.

Most primate species have color vision—a useful ability if, like many primates, you feed on fruit. Hearing and smell are less well developed (exceptions include nocturnal primates). Other primate features include hands and feet, not paws, in almost all cases, with digits tipped with nails, not claws. In most species an opposable thumb, combined with forward-facing eyes that enable accurate focus, gives the ability to hold and manipulate objects.

Primate species vary greatly in behavior and diet, as well as size. Most species live in trees in forests or woodland, but some have become savanna dwellers. A few eke out a living in arid lands, and one or two species (our own included) swim. Many primates eat leaves and fruit, others sip nectar and sap; some catch insects, lizards and other small animals; others hunt and kill quite large mammals, including other primates. Some species

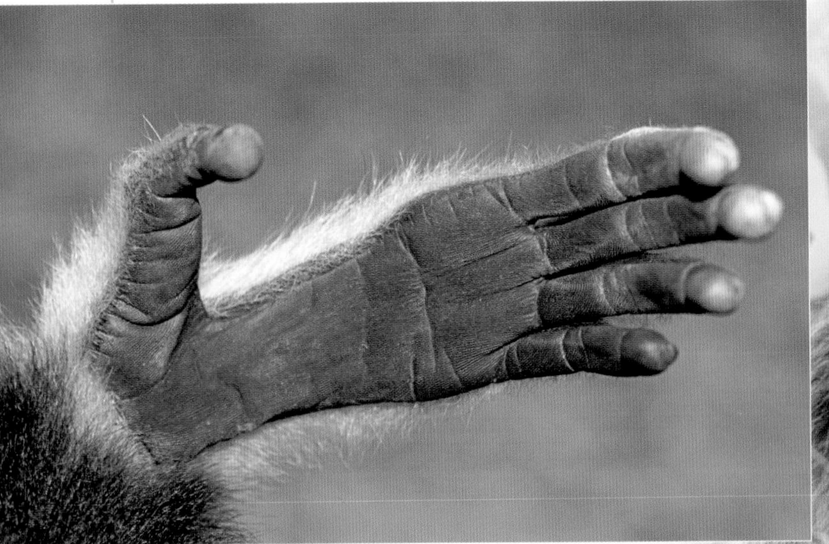

Above An opposable thumb, such as that of the White-handed Gibbon above, enables primates to grasp branches. Some species have used this characteristic to hold tools. Gibbon fingers are also very long, which allows them to hook easily onto and off branches when swinging along (brachiating).

are very specialized and are found only in particular habitats, but others are generalists that have the ability to adapt to different conditions—our own species being the most successful example. Indeed, such is our success that our expansion into every habitat on the planet is now threatening the future of all other primates.

Primate social structures

Primates are known for their complex social systems (see pages 22–23) and their intelligence, which enables individuals to learn from experience. Not all primates live in groups, but all are acutely aware of their neighbors and adapt their behavior according to their memory of the last encounter. The long period of maternal dependency, or childhood, also allows the next generation to learn from their parents' experience. In long-lived species that live in permanent social groups, we see the emergence of the role of the grandparents, whose lifelong experience influences their grandchildren. In this way, some learned behavior patterns are passed on, leading to the emergence of culture. This is most obvious in humans and the other great apes, where the ability to make and use tools has enabled us to solve problems and access food or water that might otherwise be unavailable.

THE SEVEN AGES OF PRIMATES

First, the infant in the mother's arms—baby primates are totally dependent on their mother for food, warmth, security and transport.

Next is the juvenile primate—playful and imitative, learning what to eat and where to find it and the rules of its species' society.

Third comes the adolescent and the beginnings of a sexual awakening.

Fourth is the new parent, whose youthful urge to play is tempered by a new responsibility.

Fifth is the experienced parent, dealing with offspring of different ages and secure in the social hierarchy.

The sixth age marks a shift to become an older, wiser member of the social group—a grandparent who will keep order and settle disputes between younger members of the troop.

Finally, for the few who live to their species' natural lifespan, comes old age and disease, but in most cases predators bring the final act to a close before then.

Below Young Chimpanzees spend the first few years of life learning about the world from the security of their mother's warm embrace.

Global Primate Distribution

Primates are found on virtually every area of land on the planet, but if we remove humans from the picture their distribution is more limited. Nonhuman primates are mostly found in tropical and subtropical forests and woodland. A few species have adapted to savanna and montane habitats, but even there primates are usually associated with trees and bushes. The two main limiting factors are rainfall—or the lack of it—and the need for a year-round food supply. Human ingenuity has overcome these, enabling us to colonize the driest deserts and the cold Arctic regions, but nonhuman primates cannot survive the long winters at latitudes higher than about 40 degrees.

MADAGASCAR

Prosimian primates in Madagascar (off the coast of East Africa) have evolved into a wide range of specialized species, due to lack of competition from apes and monkeys. More than 60 living species of lemur and their kin have been described, including the highly social Ring-tailed Lemur (above).

THE AMERICAS

Surviving patches of rainforest in the Yucatan Peninsula of southern Mexico represent the northernmost outpost of nonhuman primates in the Americas. The region is home to spider monkeys and the Venezuelan Red Howler Monkey, shown here making its famously loud call. The call of howler monkeys is a deafening roar that is louder than the call of any terrestrial animal and exceeded in volume in the animal kingdom only by the songs of baleen whales.

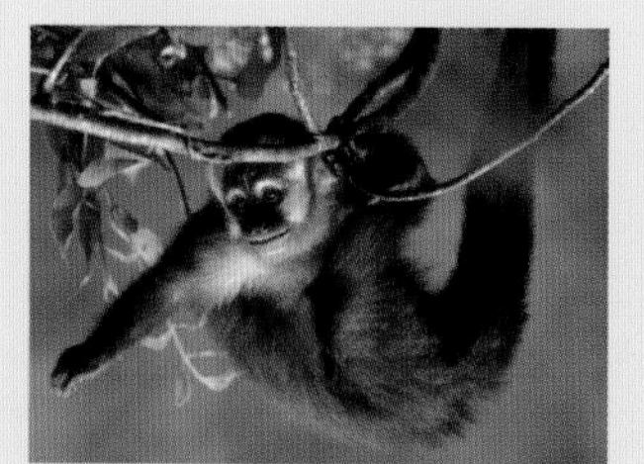

Capuchins are widely distributed in Central and South America. They are noted for their intelligence, which once made them popular in Europe as organ grinders' monkeys and as pets. Primates do not make good domestic companions, however: the risk of disease and injuries from bites, conservation considerations and a greater awareness of their needs has led most people to conclude that primates are better left in their natural habitats.

Marmosets and tamarins have diversified into some 60 species and subspecies across Central America and in South America east of the Andes. They are all small, weighing under 2.2 lb. (1 kg), and have clawlike nails on all digits except the big toe, which has a flat nail like those of other primates. These fascinating creatures have evolved in a huge variety of colors and sport flamboyant tufts of hair on the ears, mustache and mane.

EUROPE

Once found throughout southern Europe, the Barbary Macaque is now limited to cedar forests in Morocco and Algeria, as well as the famous introduced colony in Gibraltar, popularly known as the "Apes on the Rock." Not true apes, these tailless monkeys are the only nonhuman primates living wild in Europe today.

ASIA

The Japanese Macaque, or Snow Monkey, has a thick, shaggy coat and is the most northerly nonhuman primate. The species is found on all the main islands of Japan except Hokkaido, where in winter it is too cold for it to survive.

In India and Sri Lanka, due to respect for the Hindu monkey deity Hanuman, monkeys such as these Rhesus Macaques are tolerated even in towns and cities. They help themselves to human food supplies, such as garden vegetables and fruit from market stalls, but also receive deliberate offerings from devotees.

The tiny Philippine Tarsier is among the most easterly tropical primates. This species is active at night, clinging to and leaping between vertical branches in search of prey—mainly insects and lizards. Like all forest-dwelling species in Southeast Asia, tarsiers face habitat fragmentation and loss due to logging, fires and conversion of forests to agriculture, especially oil palm plantations.

AFRICA

Apart from humans, Chimpanzees are the most widely distributed great ape. Until recently they were found in forests and woodland in 26 African countries, but they are now extinct in five of these. Numbers of all ape species continue to decline in almost every surviving population.

Baboons are among the most successful primates. They are adaptable omnivores, able to live in habitats ranging from cold montane forests to hot, dry savannas. Their ability to outwit farmers makes baboons serious crop pests in many parts of Africa.

Pottos are slow-moving nocturnal primates with a currylike odor and strange bony spines protruding from the back of their neck. They clamber around trees and bushes across Africa's rainforest belt, feeding on fruit, gums and insects; two-thirds of the insects eaten are ants.

Diet and Habitat

With their superb climbing abilities, primates are able to access food sources high in the treetops, where most other mammals cannot venture. Some specialize in certain foods, such as a particular insect, but many eat a wide variety of food types. Nearly all primates eat at least some fruit, and many also eat meat. In a few cases, their diet includes other primates.

Left Great apes, such as this Sumatran Orangutan, build a new nest every night. Nest-building changes the structure of the canopy by pruning the branches, which stimulates growth and creates temporary light gaps. Chimpanzees nest lower down than orangutans, as do most gorillas; silverbacks (adult males), however, nest on the ground.

Imagine the tropical rainforest as a massive department store, with different foods on different levels for foraging primates. On the first floor you can find plant stems, bark and nutritious termites and ants; there are also ripe fruits that have fallen from the upper levels. However, these foods are available to other ground-dwelling species too, so the primates at this level have to compete by being larger—they include gorillas, Chimpanzees and Mandrills. These primates sometimes dig down into the "basement" for roots, but because the ground floor is patrolled by primate predators such as big cats, most other primates choose to spend more of their time up in the trees.

Shrub layer

The next floor up in the rainforest department store is known as the shrub layer; it offers bark, sap, fruit-bearing lianas (woody vines that grow from the ground and twist around trees), and leaves of shrubs and ferns that grow directly on low branches. The forest architecture here is dominated by branchless vertical trunks. Some fruits, including figs and cocoa pods, grow at this level; unlike trees in colder areas, many tropical trees produce fruits that grow from the main stem. Insects also feed on the fruits, and lizards and birds feed on the insects. Some insect-eating primates, including prosimians such as lorises, Pottos and bushbabies, spend nearly all their time at this very dark level, up to 30 ft. (10 m) or so above the ground.

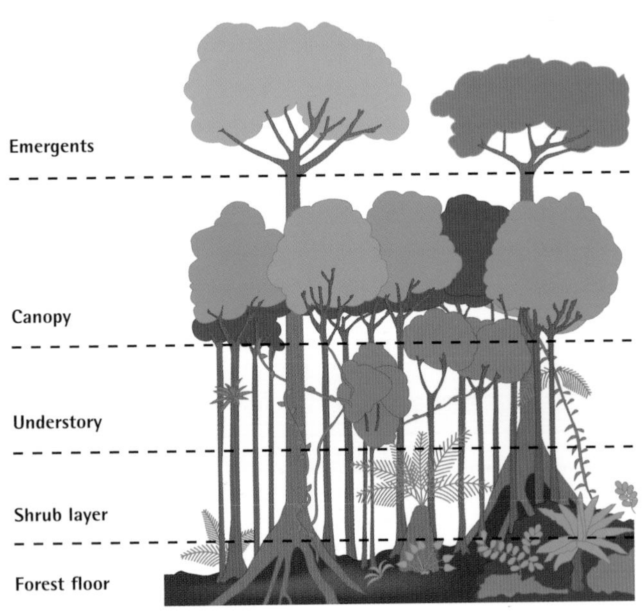

Above Tropical rainforests are more diverse than any other ecosystem, with different species at different levels, from the ground to the crowns of tall trees.

Understory and canopy

Many monkey species run along the intertwined branches of the next layer, the understory, which lies beneath the forest roof, or canopy, at about 30–80 ft. (10–25 m). The larger trunks here are still rising straight upward, but more branches come off at angles, and there are lianas and creepers climbing up toward the light, as well as the roots of strangler figs heading down toward the soil.

The canopy, beyond which only the tallest trees project, absorbs most of the energy for all the forest layers. In the canopy the foliage is bathed in sunlight and photosynthesis is at its most efficient. Some flowers and fruits can be found at the branch tips, but at 80–160 ft. (25–50 m) above ground the canopy is hazardous: misjudge the strength of a branch and you will plummet to your death. Nevertheless, gibbons and spider monkeys swing beneath the branches, picking off fruit, flowers and buds, and leaf-eating monkeys sit on the branches plucking fresh young leaves.

Below Monkeys such as this Saddleback Tamarin in Peru eat flowers to obtain nectar, a high-energy food source, and in doing so may assist pollination.

Outside the forest

Some primates exploit different habitats outside the forest. Gelada Baboons in Ethiopia live on spectacular cliffs; lowland gorillas wade in swamps to find water plants; and Patas Monkeys have evolved long legs for running fast across open grassland. But, wherever they are, trees remain important, and it is primates' climbing ability that gives them, literally, the upper hand.

WHO EATS WHAT WHERE?

Food	Africa	Asia	The Americas
Leaves in the canopy	Colobus monkeys	Langurs Proboscis Monkeys	Howler monkeys
Fruits	Chimpanzees Guenons Mandrills Mangabeys Angwantibos	Gibbons Macaques Orangutans	Capuchins Spider monkeys Titi monkeys
Sap and gum	Some lemurs Pottos	Pygmy Slow Lorises	Tamarins Marmosets
Stems and roots	Mountain Gorillas Baboons	Tibetan Macaques	
Grains and seeds	Patas Monkeys Baboons	Japanese Macaques	Bearded sakis
Bamboo specialists	Eastern Gorilla Golden Guenons Bamboo lemurs		
Insects and other invertebrates	Bushbabies Pottos Aye-ayes (grubs)	Lorises Macaques Tarsiers	Capuchins Squirrel monkeys Titi monkeys
Frogs and crabs	Baboons	Long-tailed Macaques	
Hunters of mammals	Chimpanzees Baboons	Macaques	Brown-capped Capuchins

Gardeners of the Forest

If primates are to survive in the wild, it is essential to protect their habitats. What is less obvious is the converse—if primate habitats are to survive, it is essential to protect the primates. This is because of the dynamic interplay between animals and plants and their mutual dependence. Just as a gardener manages a garden by sowing seeds, pruning, lopping branches and controlling pests, so primates perform these tasks for the forests in which they live.

When a leaf-eating monkey takes the nutritious bud and young leaves from the tip of a twig, the plant benefits. The growing tip had been suppressing the growth of lateral buds behind it along the branch—a process called apical dominance. When the tip is removed, the side buds grow faster, resulting in bushy growth. The process also creates small, temporary gaps that allow sunlight to reach leaves that were previously shaded by those just eaten.

The creation of larger light gaps in the forest canopy also results from branch-breaking, whether accidental (for example, when a leaping primate misjudges the strength of a branch) or deliberate (as when branches are cracked in social displays or by great apes building their nests). For a wood-boring beetle or termite, an old ape nest must represent a bonanza of dead wood. The holes left by the beetle larvae allow fungal spores to enter, and the decaying nest gradually returns its nutrients to the forest floor.

Fertilizer, seed dispersal and germination

Primates also fertilize the forest floor with their dung, but there is another even more important fecal function when the primates have been eating fruit: seed dispersal. The sweet, fleshy pericarp—the tissue surrounding the seed—that fruit-eating animals love is, from the plant's point of view, a lure. It tempts animals to swallow the seeds contained within it and thereby act as seed-dispersal agents. If you visit primate habitats, try teasing apart primate droppings and you may learn a lot—especially the type of fruit the monkey or ape has been eating.

Germination trials of seeds collected from such droppings show that they grow much better than seeds collected from beneath the parent plant; more seeds germinate and seedling survival is higher. This is because these seeds have evolved to survive passage through an animal's digestive system, and are so tough that if they don't get eaten, water is not able to penetrate the seed wall and germination does not occur. There are clear advantages for a seed to be planted in a neat package of fertilizer far from the parent plant—dung is packed with nutrients, and primates are likely to have traveled some distance in the time it takes for the seed to pass through the digestive tract. Thus, future generations of fruit trees depend on the continued presence of seed-dispersal agents. And while small seeds such as those in figs can also be dispersed by birds, larger ones, such as those in durian fruits in Southeast Asia, need larger animals such as orangutans to do the job.

Population control

Primates garden the forest in other ways too, such as controlling the numbers of insects that damage foliage. A glut of caterpillars on one plant might remove all of its leaves, so caterpillar-eating primates control their numbers and benefit the plant. Their dextrous ability to deal with the caterpillars' irritant hairs by wiping them off on a branch gives primates an advantage over other insect-eating animals that are deterred by such defenses.

Left Gorillas use their powerful jaws to crack open branches to extract the soft edible center, or pith, of certain food plants. By breaking saplings, they maintain light gaps, which encourages growth of herbaceous plants at ground level.

Right Adult male orangutans can eat fruits with big seeds, and so act as seed-dispersal agents for plant species whose seeds are too big for birds or monkeys.

Above Plants have evolved sweet, fleshy fruits to encourage animals to eat their seeds (1), which are then deposited far from the parent plant in a convenient package of manure (2). The seeds germinate readily (3), eventually growing into mature trees that provide more food for primates (4).

Left This tasty Myrianthus fruit lies on the forest floor in Central Africa. The fruit is enjoyed by apes, elephants and humans.

Above These Myrianthus seeds lying in gorilla dung on the forest floor will germinate more successfully than those that fall directly from the tree.

Gorillas had not been seen to use tools in the wild until primatologist Thomas Breuer took this series of photos while observing Western Lowland Gorillas in Mbeli Bai, northern Congo, in 2005. Leah, an adult female, was seen peering into a pool. She then waded in, but when the water reached her waist, she grabbed and broke off a stick. Leah first used the stick to test the depth of the water in front of her, or to make sure the mud was firm enough for her to walk on.

Primate Cultures

Say the word "culture" and many people will think of opera, ballet or art. In a wider sense, culture may be taken to mean the traditions, music and crafts of different human societies. It is only in the past 20 years or so that the idea of culture in nonhuman primates has gained acceptance, and there are still some academics who question it today. As more evidence emerges from field studies, however, the case for nonhuman culture becomes more compelling, and most researchers now consider that human culture differs only by degree from that of other primate species.

One day in 1953, a young female Japanese Macaque had the idea of washing the sand off her sweet potatoes in the sea before eating them, no doubt to make them taste better. She had been christened Imo by a team of scientists that provided potatoes on the beach so they could study her troop in the open. The zoologists were astonished by Imo's innovation. They had been studying the monkeys for five years by then and had never seen anything like this. Imo's method of preparing grit-free dinner was copied by her playmate, then her mother, and within a few years, all but the old males in the troop were doing it. Today, new infants still learn it from their mothers, and it has became a tradition found only in that particular troop of macaques.

Clever monkey

Imo's innovations went beyond potato washing—the scientists also fed the monkeys with grains of wheat, which tended to get mixed up with the sand. Imo discovered that if a handful of sand and wheat was dropped into seawater, the sand sank and the wheat floated, which both disposed of the grit and made the grains much less fiddly to eat. This clever discovery was also copied by others in the troop. Not only did washing the sweet potatoes and wheat get rid of the gritty sand, but the salt water probably improved the taste too. Not all examples of macaque culture involve food, however. Another tradition of playing with stones began in a troop of macaques near Kyoto, Japan, in 1979; four years later, half of the troop were tapping stones together, despite it having no apparent benefit beyond the pleasure of play.

Adjacent cultures

Cultural differences can sometimes be seen between even close neighbors. One of the favorite foods of Mountain Gorillas is the stinging nettle; young gorillas closely watch their parents to learn the least painful method of preparing nettles. One young female known as Picasso grew up in a group whose home range did not include nettle beds. When she reached maturity, she joined a neighboring group that did have nettles, but she ignored this important food source, even though others in her new group ate it. She had learned her food-processing skills as an infant and didn't seem inclined to change old habits. Or it may be that because gorillas feeding in dense vegetation are usually out of sight of other adults, there may be few opportunities for learning new skills. Either way, when Picasso had a baby, it too failed to learn the nettle-feeding skill—an example of a family tradition.

When first reported, these kinds of behaviors led primatologists to begin a discussion about what constitutes culture. Many researchers now agree that culture in this context can be defined as a behavior that is specific to members of one group or population, and that is transmitted socially by one animal learning it from another. This definition can also apply to any human culture. It suggests that all of our complex inventions and traditions stem from an ability that is shared by other species, and so culture is not uniquely human.

Facing page An adult wild Brown-capped Capuchin uses a heavy rock to crack palm nuts, watched closely by a juvenile in Cerrado savanna habitat, Brazil. Scientists were amazed to observe this behavior, which is exhibited by only a few populations in Cerrado and therefore appears to be an example of monkey culture.

Next, she used the branch as a walking stick to help balance herself while wading into the pool—behavior never seen before in a gorilla. She waded 25–30 ft. (8–10 m) further into the pool using the walking stick, then left it in the pool and rejoined her infant who was crying on the bank. It is not known yet whether this is a one-off invention, or a local culture, but a month later another unrelated female used a stick for support, and research continues at the site.

Social Structures

Watching primates interact with one another is immediately fascinating. If you study them long enough to recognize individuals, it becomes even more interesting; it becomes a question of who is doing what to whom, and why? Social behavior is the focus of many primate studies, partly for the intrinsic interest in learning about the species in question, but also because an understanding of other social systems can help us to understand our own better.

Primates are intelligent, social mammals with as much diversity in their social structures as in their appearance. The way in which their societies have evolved reflects the constraints of their different habitats, the threats from predators and competition from other species, as well as from other members of the same species. Some primates spend most of the time alone; some live in monogamous pairs; some form large, permanent social groups—either single-male/multifemale groups or multimale/multifemale groups; and some live in fluid communities, spending different times with different members of their social group for reasons it can be difficult for a human observer to work out. And some primate species have yet to be studied, so very little information is available about how they live.

There are both advantages and disadvantages to group living. With more pairs of eyes and ears alert to danger, each group member can spend more time feeding or relaxing, safe in the

Facing page Gelada Baboons form the largest social groups of all wild primates. This group is feeding, each individual monitoring relationships and pecking order.

knowledge that someone else will raise the alarm if a predator appears. Sometimes different species of monkeys will feed together in the forest canopy, each helping to keep a lookout for monkey-eating eagles. By working together, even large predators can be chased away—for example, baboons are able to see off a Leopard by ganging up on it, whereas if the Leopard had sneaked up on a lone baboon, the victim would not have stood a chance.

The more members there are in a group, however, the more mouths there are to feed, so group size is often determined by the availability of food. In a group, you not only have to defend your food resources from other groups (which may lead to a system of defending exclusive territories), but you may also have to compete for limited food with members of your own group, which can lead to intragroup aggression. The ways in which such conflicts are resolved form the key to group living. Aggression is usually followed sooner or later by appeasement and reconciliation, because the long-term benefits of living together outweigh the short-term disadvantages of having to share.

Parenting

Primates have a long period of maternal dependence—in other words, their childhood—during which time they must learn the rules of their society, where to find food and how to survive. This imparting of information usually falls to the mother, who provides transport, warmth and security, so the mother–offspring bond is the strongest of all bonds within primate societies and may last for life. In some primate species, such as marmosets, gibbons and gorillas, the father also plays an important contributing role in caring for the young; in others, such as Pottos and orangutans, childcare is exclusively the mother's responsibility.

Much of primates' social learning takes place while they play with each other. Knowing how to interact safely with other members of the group, or with neighbors, is important for social success and ultimately leads to reproductive success. The larger the social group, the more complex this process can be—particularly when the relative status of a child's parents may influence the outcome of interactions between the child and its peers. And with higher intelligence, these complex social interactions may involve political alliances, careful planning and even deception in order to achieve an advantage over a rival. Clearly, much of what we consider to be human social behavior is in fact shared by many of our primate relatives.

A MAN'S WORLD?

When two animals meet, they quickly establish which of them is stronger or more confident than the other (incidentally, this is not always the same thing). Whether this is settled by some kind of signal, display or physical struggle, the dominant individual will then have right of way and gain first access to food or a mate. The submissive animal will give way. Animals such as primates, with the capacity to remember this information the next time they meet, soon learn where they are in the local dominance hierarchy. Another term for dominance hierarchy, first described in relation to domestic chickens, is the pecking order.

In many species, the struggle for dominance through natural selection has led to differences in size or shape between the sexes. This is known as sexual dimorphism. In most primate species, for example, males are usually bigger than females, and this phenomenon can be seen in gorillas and baboons. The most dominant individual in a group is known as the alpha male. In some species, however, leadership is decided by other qualities and it may be an alpha female that is in charge, as happens with Bonobos and most lemurs.

Above This adult Siamang Gibbon is vocalizing with its resonating throat sac fully inflated; the females give a series of barks to complement the male's screams in their daily morning duet. Siamangs are sociable animals. They are protective of one another and huddle together in small groups when they sleep.

Communication

Primates are consummate communicators. The more complex a social group or network, the more important it is for each individual to know what other members or neighbors are doing. Whether by scent, sight, sound or touch, communicating information about where you are, what you are doing, how you are feeling and whether you are friend, foe or sexually receptive is essential for success in primate society.

Watch any group of primates acting naturally, such as a troop of monkeys or lemurs, or a family of apes or humans, and it is immediately obvious that information is passing back and forth between the individuals. Most animal communication is to do with announcing the mood, social status or sexual state of the communicator. Small facial movements can convey a subtle threat or invitation to a target individual, while whole-body stance or movement, perhaps emphasized with a vocalization or odor, can make the same signal very obvious to all observers.

Visual signals are important in many primate species, and some have evolved physical features to emphasize such signs. Patches of colored skin on the face or genitals, tufts of hair on the ears, eyebrows or lips, and even bony or fleshy protruberances—these all say something about the owner. Human eyebrows are a good example: when raised, they express surprise; angled, they express anguish or fear; lowered over a direct stare, they convey a threat. Even though other primates do not have such prominent eyebrows

Above This Ring-tailed Lemur carrying an infant is scent-marking a sapling. As well as providing a visual display, the tail wafts chemical messages to other lemurs, informing them of the identity and sexual condition of the signaler.

Above Two young Proboscis Monkeys make faces at each other in play. In many species of primates, the "play face" has an open-mouth but with lips covering most of the teeth, as distinct from an aggressive face in which they are bared.

Excited

Frightened

Playful

Interested

Above Many primates have hairless faces that display for others every blush and expression, as hundreds of muscles and subcutaneous blood vessels respond to and reveal their emotions. These images show typical Chimpanzee expressions.

on bare facial skin, these same signals are common to many species, so a human observer can often, but not always, correctly interpret the facial expressions of other primates based on our expressions. Most people see a performing Chimpanzee's grin as a smile, but sadly, many trainers discipline their animals with violence to make them perform on command, so the "smile" is actually a fear grin.

Common signals

Yawning is a commonly seen signal that also exposes the teeth, although the reason for it is still open to debate. A yawn in humans is usually taken to mean either tiredness or boredom, but in other primates it can also indicate nervousness or stress. Recent research suggests that it might have a brain-cooling function, permitting clear thinking—which would certainly be useful in situations that might result in someone feeling nervous. But the effect of a yawn is surely to reveal an animal's most potent weapons—the canine teeth—so perhaps it is just a matter of reminding all those around you of your potential to bite, before going to sleep, when waking up or just feeling anxious, to discourage others from bothering you.

Some primates flip the upper lip up to expose the teeth as a more overt threat. Conversely, lip-smacking is used by many species to say, "I'd like to groom you," either inviting a friendly interaction or attempting to defuse an aggressive encounter.

All primates use sound for communication, both for intimate signals at close quarters and long-distance messages: loud territorial calls, such as orangutan "long calls," gibbons singing and howler monkeys bellowing, announce ownership of a patch of habitat; soft grunts may reassure and denote affection; squeaks and hoots can convey excitement. Although there are many different vocalizations for different occasions, nonhuman primates do not have a language as such. In some cases, however, vocalizations may convey surprisingly precise information. Researchers studying Vervet Monkeys tried playing back recordings of different alarm calls and found that the monkeys had different sounds for a snake, bird of prey and Leopard, and they reacted differently according to where the threat seemed to come from (see page 124).

Above For many primates, grooming, as in this group of Bonobos, serves both hygienic and social functions, and is a sign of friendship or a family bond, rather like wordless, tactile conversation.

Studies of captive apes have revealed that, although they do not possess the vocal apparatus for human speech, their brains certainly do have the capacity for language. All the great apes have shown that they can learn sign language or use abstract symbols to construct simple sentences with an understanding of syntax. This suggests that the part of the brain that deals with names of objects and individuals evolved in our common ancestor, before humans and the other great apes split on their separate evolutionary branches. It also implies that a more careful appraisal of ape vocalizations in the wild might reveal that there is more to those subtly different grunts than meets the untutored human ear.

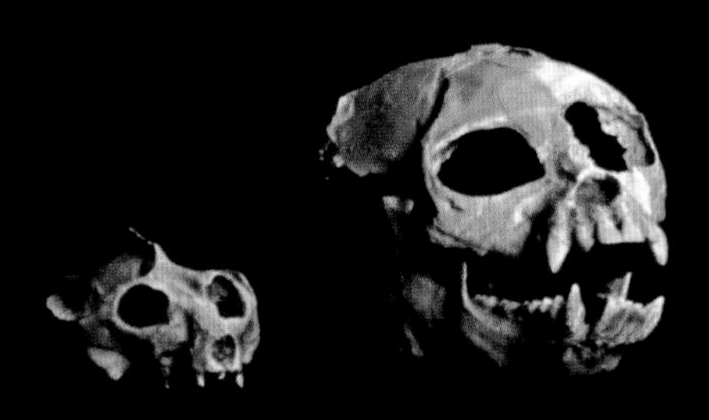
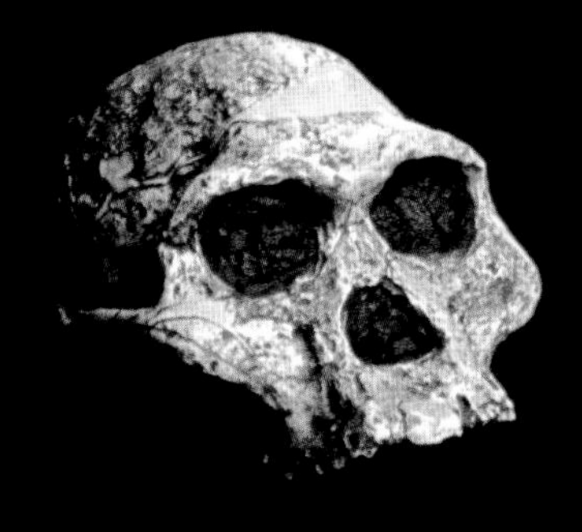
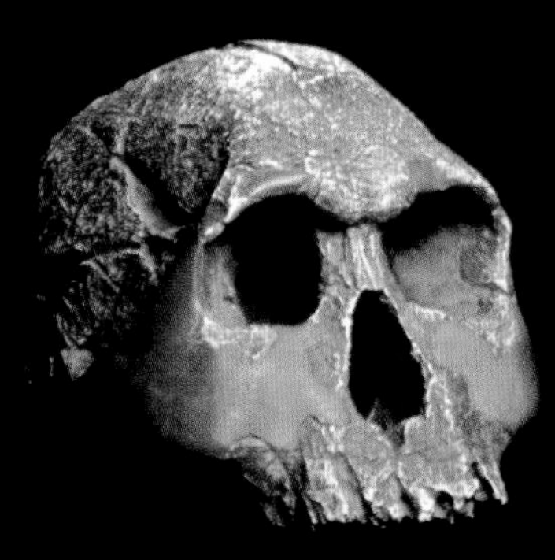
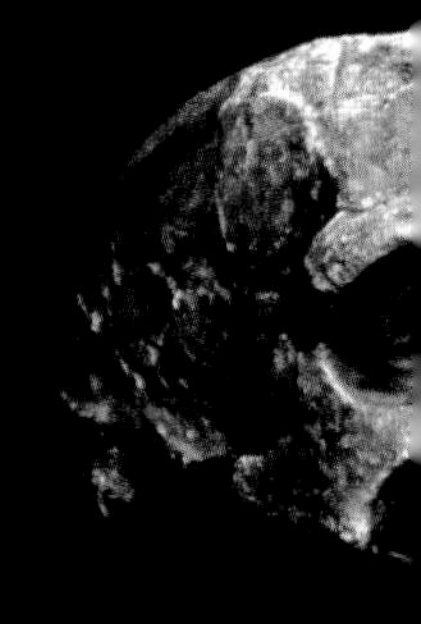

Darwin's Big Idea

During the 19th and early 20th centuries, in the aftermath of Darwin's revelations, ideas about primate origins were based mainly on comparative morphology—comparing the anatomy of different species. But the discovery of DNA in 1953 cast an entirely new light on the mechanism of evolution. It led to a reinterpretation of the relationships between species and forever changed humankind's understanding of its place within the animal world.

In the 18th century, western naturalists had seen too few specimens to gain a clear understanding of primate diversity, but throughout the 19th century, explorers and collectors filled up museums with specimens. Although this improved anatomical knowledge, most zoology at the time was still based on the study of dead animals, spiced up with anecdotal accounts of behavior—often the behavior of distressed animals under attack by the collector! Eminent anatomists would add drama to dry scientific papers by recounting stories with little or no evidence. For example, the British biologist Professor Richard Owen (1804–92), then Superintendent of the British Museum's natural history department, studied the first gorilla specimens to be seen in the United Kingdom and published accurate skeletal data in serious scientific journals. But he also wrote: "[Natives] when stealing through the shades of the tropical forest become sometimes aware of the proximity of one of these frightfully formidable apes by the sudden disappearance of one of their companions, who is hoisted up into the tree, uttering, perhaps, a short choking cry. In a few minutes he falls to the ground a strangled corpse." Why gorillas would behave in such an unaccountable way was not explained.

In the 1850s, when the earliest stories of live gorillas first began causing a stir in Victorian drawing rooms, parts of a strange skeleton were discovered in Germany's Neander Valley. These bones became the type specimen of "Neanderthal Man" (*Homo neanderthalensis*).

Above This 19th-century artwork by the German biologist Ernst Haeckel depicts the skeletons of five primates: from right to left, a gibbon, an orangutan, a Chimpanzee, a gorilla and a human. Haeckel mistakenly believed that gibbons were close relatives of humans, because they are the only apes that habitually walk upright on two legs (when not swinging through the trees).

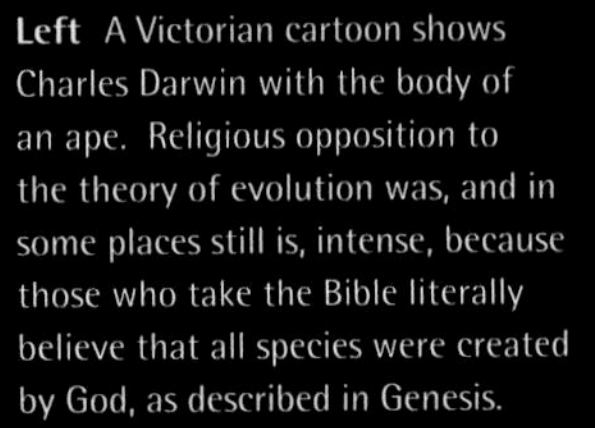
Left A Victorian cartoon shows Charles Darwin with the body of an ape. Religious opposition to the theory of evolution was, and in some places still is, intense, because those who take the Bible literally believe that all species were created by God, as described in Genesis.

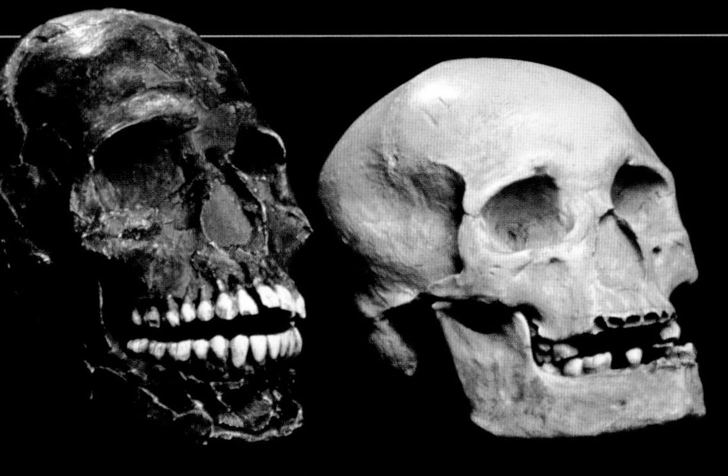

Above These seven skulls belong to ancestors and relatives of modern humans. From left to right the skulls are: *Adapis* (a lemur-like animal that lived around 50 million years ago); *Proconsul* (a primate from 23–15 mya); *Australopithecus africanus* (3–1.8 mya); *Homo habilis* (2.1–1.6 mya); *Homo erectus* (1.8–0.3 mya); a modern human (*Homo sapiens sapiens*) from around 92,000 years ago; and a French Cro-Magnon human from about 22,000 years ago.

Above People say that apes "look human" but it is more accurate to say that humans and apes have many features and expressions in common. This similarity supports the Darwinian view that we share a common ancestor.

KEY DATES IN PRIMATOLOGY

1735 Swedish naturalist Carl Linnaeus publishes the first edition of his classification of living things, the *Systema Naturae*. Subsequent editions recognize that humans fall within the zoological order he named Primates.

1783 James Hutton, a British geologist, publishes his *Theory of the Earth*, which states that fossils represent the remains of animals and plants, some now extinct.

1847 After centuries of unverified travelers' tales, the gorilla is finally described by western science as "a new species of orang from the Gaboon River."

1856 A partial skeleton of "Neanderthal Man" is found at the Feldhofer Cave in the Neander Valley, western Germany.

1859 Charles Darwin publishes *On the Origin of Species*, presenting evidence for evolution by natural selection.

1871 Darwin fuels the "Origin" controversy by publishing *The Descent of Man*, which includes detailed evidence of human evolution, causing further outrage in the Christian Church.

1917 Wolfgang Kohler publishes the results of his intelligence tests on captive Chimpanzees, showing their tool-using and planning abilities.

1930 Dr. Robert Yerkes of Yale University opens a primate research facility in Florida, the first in North America.

1953 Scientists James Watson and Francis Crick publish their work on the double-helix structure of DNA—the code of life.

1960 Jane Goodall begins her long-term study of Chimpanzees at Gombe on Lake Tanganyika, Tanzania.

1984 Charles Sibley and Jon Edward Alquist publish the results of a DNA–DNA hybridization study showing that humans share 98.4 percent of their DNA with Chimpanzees, 97.7 percent with gorillas and 96.4 percent with orangutans.

1997 Simon Eastel and a team at the National University in Canberra, Australia, announce that their research indicates humans and African apes shared a common ancestor as recently as 3.6–4 million years ago, and that this ancestor must have walked upright.

2000 Scientists predict that, at the current rate of decline, most populations of gorillas, Chimpanzees, Bonobos and orangutans will be extinct within 20 to 30 years.

2005 The United Nations brokers the Kinshasa Declaration on Great Apes, in which governments agree to a global strategy for their survival.

2007 The United Nations Convention on Migratory Species negotiates a treaty between the 10 countries with gorilla populations to ensure the survival of these primates.

Often thought of as brutish and unintelligent, Neanderthals in fact had larger brains than modern humans, and their remains have since been found together with evidence of a rich culture. Debate still rages over whether modern humans exterminated the Neanderthals or interbred with them and assimilated their genes into the population. Studies of the Neanderthal DNA found in fossil bones may finally answer this question.

The heady mix of Darwin's ideas, more detailed observation of the great apes and discoveries of new evidence of fossil ancestors caused 19th-century scientists to speculate on the existence of a "missing link"—an intermediate species somewhere between humans and nonhuman apes that would prove Darwin's controversial theory of evolution and that might also shed light on the origins of our own species, *Homo sapiens*.

THE PRIMATE FAMILY TREE

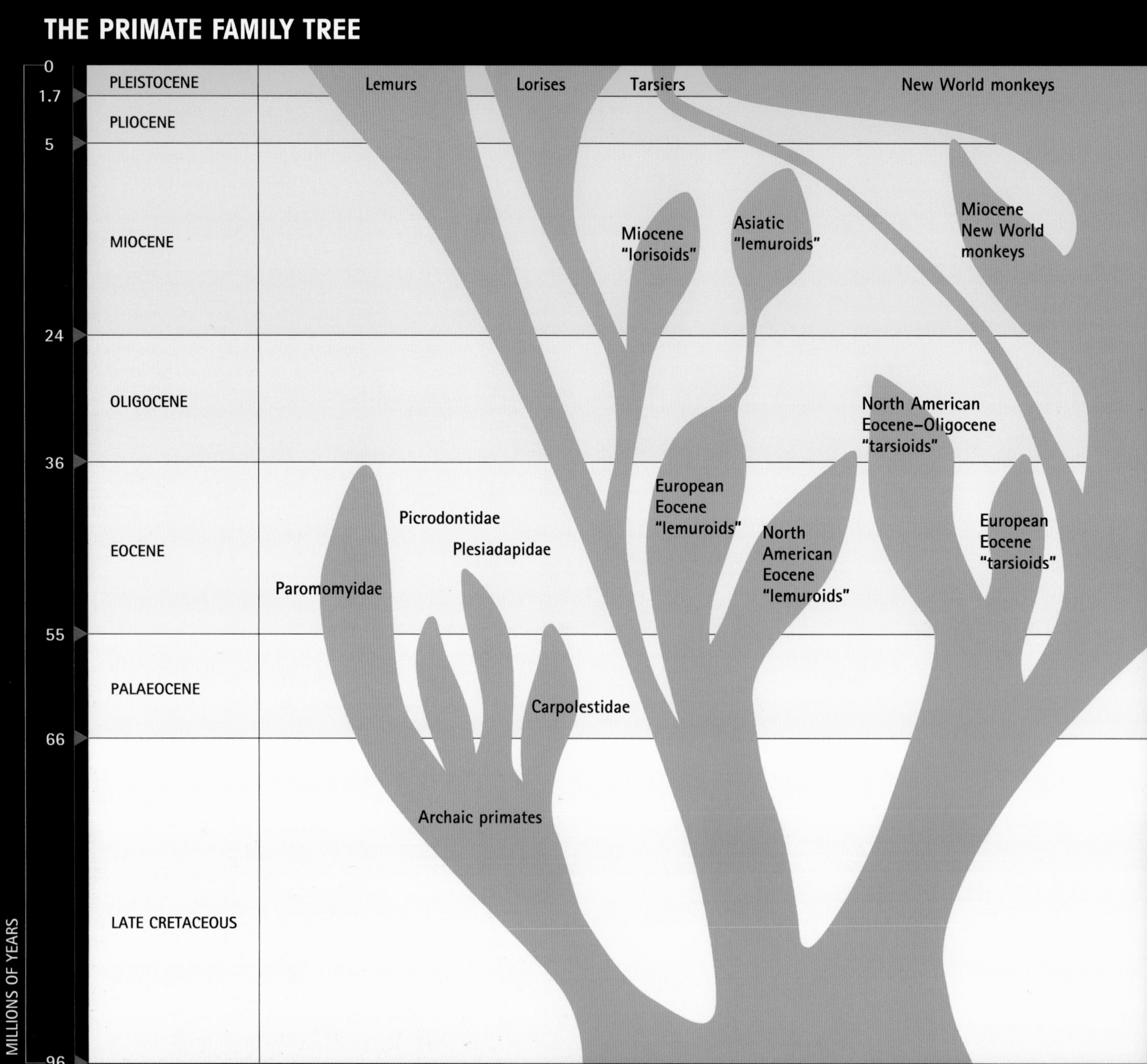

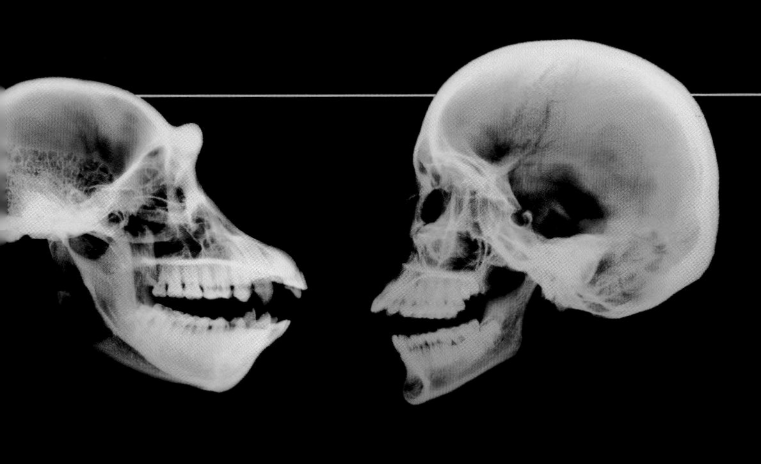

Left This false-color X-ray image shows the skulls of a Chimpanzee (left) and a human (right) viewed from the side. The Chimpanzee's cranium (the spherical part of skull that contains the brain) is much smaller than that of the human. There are prominent brow ridges above its eyes, and the large mandible bone (lower jawbone) has powerful chewing and biting muscles attached to it.

Just another primate species

By the 20th century, a growing acceptance of the evidence fo evolution by natural selection had led to a reappraisal of "man's place in nature." Instead of seeing humans as the pinnacle o creation, at the top of the ladder of life, scientists began to view humans as just another species, albeit a remarkable one with the ability to alter the very planet on which it lives.

The discovery of deoxyribonucleic acid, or DNA, in 1953 also fec this reappraisal. Once the mechanism of evolution was understoo and it was realized that human inheritance follows the same rule and uses the same chemicals as that of other animals, the logica conclusion was that Carl Linnaeus, the father of modern taxonomy had been right: we are primates. And Darwin was right too: we share a common ancestor with other primates. The big questior then became: how long ago did this last common ancestor live?

Based on fossil evidence alone, paleoanthropologists in the 1970s would have estimated that human and ape lines las branched 20 million years ago. But once methods of comparin DNA were developed in the 1980s, providing a molecular clock fo evolutionary change, the date of the last shared common ancesto dropped to 5 million or 6 million years ago. Studies of differen parts of the human and Chimpanzee genome led to debates ove the percentage of DNA we have in common with our closes relatives and, by inference, how long it took to become the specie we are today. The biggest challenge to traditional thinking cam in 1997, when an Australian team concluded that this key dat was only 3.6 million–4 million years ago—some time after biped locomotion first appears in the fossil record. If this stands up t further scrutiny, the implication is that gorillas and Chimpanzee shared a bipedal, ape-man-like ancestor with us, and then becar knuckle-walkers and returned to the forest while we spread acro the savanna and ultimately colonized every habitat on Eart Primatology seeks evidence to back up or destroy such challengi theories and helps us to know ourselves. And each new stu published confirms that many discoveries are yet to be made.

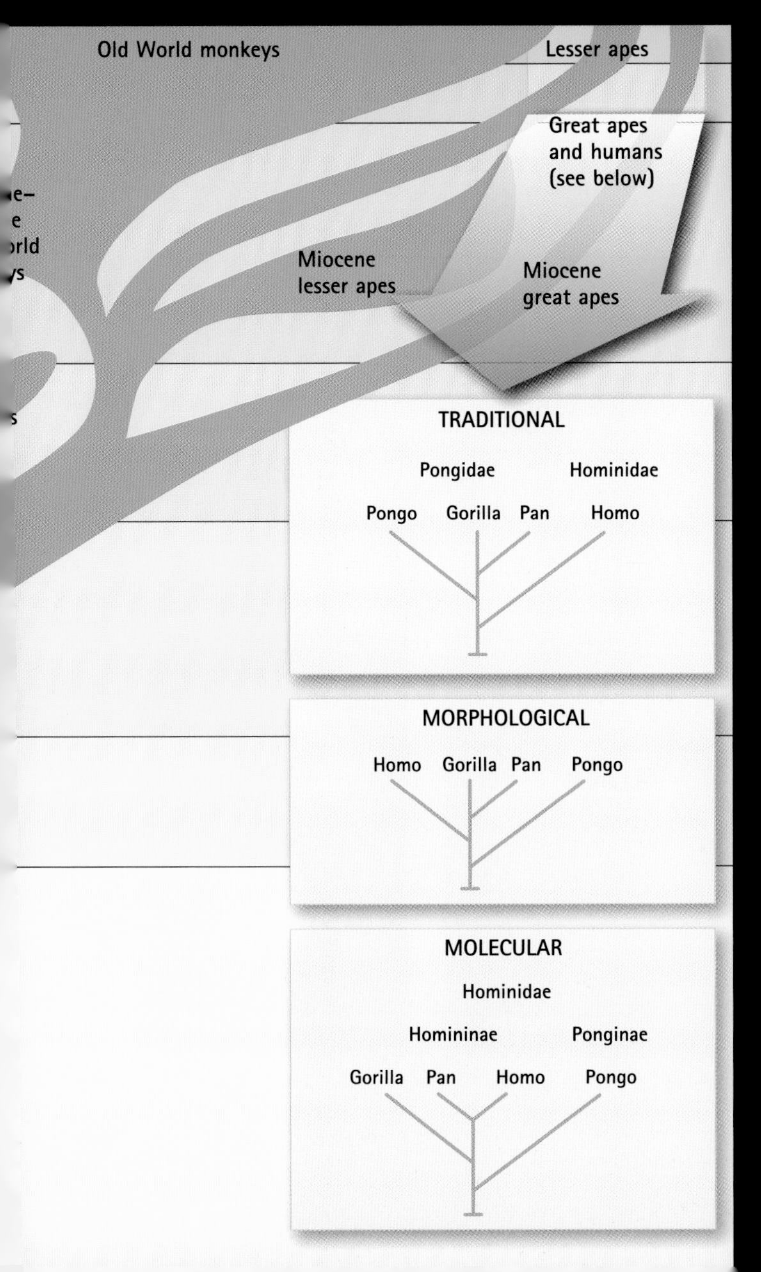

Left These three diagrams show possible family trees for the great apes and the humans, or Hominoidea. The traditional tree shows man (*Homo*) splitting early from the other great apes—orangutans (*Pongo*), gorillas (*Gorilla*) and Chimpanzees (*Pan*)—to form its own family, Hominidae. The morphological tree based on a study of anatomy, shows *Pongo*, the Asian apes, splitting first from the "trunk." The molecular tree, based on a study of DNA, shows humans, gorill and Chimpanzees forming their own subfamily of African apes, or Homininae, i which the humans and Chimpanzees are most closely related.

ORDER | SUBORDER | INFRAORDER | SUPERFAMILY

PRIMATE TAXONOMY

Taxonomy is the science of classifying living things. Scientists divide living things into five kingdoms: Monera (bacteria), Protozoa, Fungi, Plants and Animals, each of which heads a seven-tier classification system to species level. Further tiers may be added, such as infraorders or superfamilies. Primates are one Order in the Class Mammalia, in the Phylum Chordata (animals with a spinal cord). Scientists continually reinterpret the evidence for classification in the light of new research, and there is little consensus over the exact numbers of and relationships between primate species. Many taxonomists fall into one of two camps: the "lumpers" emphasize shared features, and hence tend to recognize fewer species; the "splitters" emphasize differences and tend to recognize more species.

- Primates
  - Strepsirrhini (Prosimians)
    - Lorisiformes (lorises and relatives)
      - Lorisoidea (lorises and relatives)
    - Lemuriformes (lemurs and relatives)
      - Lemuroidea (lemurs and relatives)
  - Haplorrhini (Anthropoids)
    - Tarsiiformes (tarsiers)
    - Platyrrhines (New World monkeys)
    - Catarrhines (Old World monkeys and apes)
      - Cercopithecoidea (Old World monkeys)
      - Hominoidea (apes)

FAMILY AND SUBFAMILY

Related genera (groups of species) are grouped within a zoological family, which is named by adding "idae" after the name of a genus that is judged to typify the group. Some families clearly have clusters of similar genera within them, and these are classed as subfamilies and are named by adding "inae" to one typical genus name.

SPECIATION—THE PROCESS BY WHICH NEW SPECIES FORM

If a group of animals becomes geographically isolated from its main population for many generations, its members may develop different characteristics and eventually evolve into a new species. The number of different traits, which can be measured by comparing either physical features or DNA, indicates how long ago separation occurred. In tamarin monkeys, isolated populations have evolved distinctive hair tufts and coat colors, which may be linked to mate selection and illustrate speciation well. Traits evolved during speciation often involve adaptation to local conditions.

Above The Emperor Tamarin (*Saguinus imperator*) with its distinctive mustache is from western Brazil and eastern Peru.

Above The distribution of the Golden-headed Lion Tamarin (*Leontopithecus chrysomelas*) is limited to the east coast of Brazil.

Above The Cotton-top Tamarin (*Saguinus oedipus*) is restricted to a small area of northern Colombia and Panama.

FAMILY	SUBFAMILY	GENUS	SPECIES
Lorisidae (lorises and relatives)		*Loris* (slender lorises)	2
		Nycticebus (slow lorises)	3–5
		Arctocebus (angwantibos)	2
		Perodicticus (pottos)	1
Galagidae (bushbabies)		*Euoticus* (needle-clawed bushbabies)	0–2
		Galago (lesser bushbabies)	5–7
		Galagoides (dwarf bushbabies)	0–11
		Otolemur (greater bushbabies)	0–4
Cheirogaleidae (dwarf and mouse lemurs)	Cheirogaleinae (dwarf and mouse lemurs)	*Allocebus* (hairy-eared dwarf lemurs)	1
		Cheirogaleus (dwarf lemurs)	2–7
		Microcebus (mouse lemurs)	3–8
		Mirza (greater mouse lemurs)	2
	Phanerinae (fork-crowned lemurs)	*Phaner* (fork-crowned lemurs)	1–4
Lepilemuridae (sportive lemurs)		*Lepilemur* (sportive lemurs)	7-8
Lemuridae (true lemurs)		*Eulemur* (brown lemurs)	5–11
		Lemur (ring-tailed lemurs)	1
		Varecia (ruffed lemurs)	1–2
		Hapalemur (bamboo lemurs)	3–4
Indridae (indris and relatives)		*Avahi* (woolly lemurs)	1–4
		Indri (indris)	1
		Propithecus (sifakas)	3–7
Daubentoniidae (aye-aye)		*Daubentonia* (aye-ayes)	1
Tarsiidae (tarsiers)		*Tarsius* (tarsiers)	5–8
Callitrichidae (marmosets and tamarins)		*Callibella* (dwarf marmosets)	1
		Callimico (Goeldi's marmosets)	1
		Callithrix (Atlantic marmosets)	6–11
		Mico (Amazonian marmosets)	0–13
		Leontopithecus (lion tamarins)	4
		Saguinus (tamarins)	12–17
Cebidae (New World monkeys)	Aotinae (owl monkeys)	*Aotus* (night monkeys)	5–11
	Callicebinae (titi monkeys)	*Callicebus* (titi monkeys)	13-30
	Cebinae (capuchins and squirrel monkeys)	*Cebus* (capuchins)	5-8
		Saimiri (squirrel monkeys)	5
	Pitcheciinae (sakis and uakaris)	*Cacajao* (uakaris)	3–4
		Chiropotes (bearded sakis)	2–5
		Pithecia (sakis)	7–10
	Alouatinae (howler monkeys)	*Alouatta* (howler monkeys)	8–10
	Atelinae (spider monkeys)	*Ateles* (spider monkeys)	4–7
		Brachyteles (muriquis)	2
		Lagothrix (woolly monkeys)	4–5
		Oreonax (yellow-tailed woolly monkeys)	1
Cercopithecidae (Old World monkeys)	Cercopithecinae (cheek-pouch monkeys)	*Macaca* (macaques)	21
		Papio (baboons)	1–5
		Theropithecus (geladas)	1
		Mandrillus (mandrills and drills)	2
		Cercocebus (mangabeys)	6
		Lophocebus (crested mangabeys)	3
		Rungwecebus (kipunjis)	1
		Erythrocebus (patas monkeys)	1
		Allenopithecus (Allen's swamp monkeys)	1
		Chlorocebus (vervet monkeys)	1–6
		Miopithecus (talapoin monkeys)	2
		Cercopithecus (guenons)	23-36
	Colobinae (leaf-eating monkeys)	*Colobus* (black-and-white colobus)	5
		Piliocolobus (red colobus)	0-9
		Procolobus (olive colobus)	1–2
		Presbytis (leaf monkeys and surilis)	11
		Semnopithecus (grey langurs)	7
		Trachypithecus (langurs and lutongs)	17
		Pygathrix (douc langurs)	3
		Rhinopithecus (snub-nosed monkeys)	4
		Simias (pig-tailed langurs)	1
		Nasalis (proboscis monkeys)	1
Hylobatidae (lesser apes)		*Hoolock* (hoolock gibbons)	1
		Hylobates (gibbons)	6–7
		Nomascus (crested gibbons)	4–5
		Symphalangus (siamangs)	1
Pongidae (great apes)		*Pongo* (orangutans)	2
		Gorilla (gorillas)	2
		Pan (chimpanzees and bonobos)	2
		Homo (humans)	1

ECIES AND GENUS

Biological Species Concept" defines a
es as "groups of actually or potentially
breeding natural populations that
eproductively isolated from other
groups," and this definition is widely
pted among scientists. Closely related
es are classified in the same genus
al genera), which gives them their
al name; for example, *Homo sapiens*,
r Linnaeus' binomial system.

Living with the Relatives

Main image A juvenile Rhesus Macaque in Bharatpur, India, observes how an adult responds to a *saddhu*, or holy man, who is feeding what many in India regard as sacred animals.

Facing page In this detail from a painting by Hieronymus Bosch (1485), the devil is portrayed as a monkey vying with an angel for a dying man's soul.

Friend or Foe?

Attitudes to nonhuman primates differ widely between different human cultures. They might be feared as an enemy or revered as a deity, shot and eaten by one hunter or captured and beaten by another, to be sold as a performing animal or pet. In medieval times they were used to symbolize the devil, and today their lack of sexual inhibitions causes embarrassment to prudish parents. And yet these animals are so intrinsically interesting, acting as a mirror to human behavior, that they continue to fascinate us just by being themselves.

In some parts of Africa, Chimpanzees are totem animals: their intelligence is recognized by certain tribes and hunting them is forbidden. But only a few miles away, a different tribe might hunt and eat Chimpanzees precisely because they are intelligent, believing that eating the meat imbues them with some of the properties of the animal. Gorilla meat is eaten by chiefs and served to important guests in these tribal cultures, to give them the strength and power of a silverback. The contrasting beliefs of neighboring tribes help to illustrate the complexity of human and nonhuman primate relations.

Wherever people live near other primates, a variety of myths, stories and beliefs grow up about them. Before the advent of television, people who did not live in close proximity with wild primates had even more peculiar ideas, arising from travelers' tales of strange "hairy people" and perhaps an occasional encounter with a sailor arriving home with a pet monkey or an organ grinder with a trained performer. The modern appreciation of the beauty of nonhuman primates would doubtless seem strange to anyone who grew up being told they were ugly, evil or dangerous.

Changing perceptions

Plato set the tone in the West when he wrote, "the most beautiful of apes is hideous in comparison to man and the wisest of men is an ape beside God." Later, the early Christians believed a monkey or ape was a *figura diaboli* (an image

Left Rhesus Macaques receive offerings from a holy man at Durga (Monkey) Temple, Varanasi, India. Infant monkeys learn how to behave from their elders. The behavior of these monkeys, who experience non-threatening contact with humans, is totally different from that of hunted or crop-raiding macaques, demonstrating that macaques have different cultures.

of the devil), and decided that the sexual proclivity and thieving tendencies of these animals represented the epitome of sin. By the end of the Middle Ages, primate "sinners" had instead come to be seen as fools, perhaps as a result of the trained monkeys in traveling circuses and shows. Words such as "monkey" or "ape" were now used as insults. When Victorian explorers started writing of their hunting exploits in the mid-19th century, the "monster ape" image became firmly established in popular culture, culminating in the release of films such as *King Kong* (1933). However, these western views contrast greatly with the Buddhist, Taoist and Hindu traditions of respect for monkeys. In many eastern stories, monkeys are the heroes or show the path to enlightenment.

Pest control

Whether primates are traditionally portrayed as friends or foes, the reality for many farmers and plantation owners is that they can be serious pests. Western farmers are used to insects and rodents stealing corn or vegetables, but deterring large, agile, intelligent primate pests is even more of a challenge. In Kenya, farmers next to Aberdares National Park are luckier than most: a complex, many-stranded electric fence keeps all wildlife in the park, with additional loops around the fence posts to stop baboons climbing over. As a result, local people can benefit from the supply of water flowing out of the forest and enjoy seeing animals without feeling threatened. But such fences are expensive and for most such communities vigilance is the only option. As crops ripen, people must keep watch and attempt to frighten away or kill animals attracted by the prospect of an easy meal. Where forests are replaced by large-scale plantations, workers may even be rewarded for killing crop-raiding animals—including endangered species protected by law, such as Chimpanzees in Africa and orangutans in Borneo and Sumatra. There is an urgent need for less destructive solutions, such as the Aberdares fence, to enable humans and nonhuman primates to coexist peacefully.

PRIMATES AND HUMAN MEDICINE

In the Sahel region of Africa, most people carry a good luck charm called a *giri-giri*. It consists of a small pouch on a cord, worn close to the skin, and gives the wearer confidence. Its contents are known only to the wearer and his traditional medicine practitioner, who prescribes bits of different animals and plants according to his assessment of need. This is why, if you walk through a market in Mali or neighboring countries, among the stalls of vegetables you will see a stall with rows of baboon hands, Patas Monkey skulls and assorted remains of other species. It is a traditional African pharmacy. If the market is in Brazzaville, Congo, you might see gorilla and Chimpanzee hands with bits of fingers missing—the former is prescribed for strength, the latter for cunning.

In Southeast Asia, parts of thousands of animal and plant species, including some primates, are used in traditional Chinese medicine. Slow lorises, for example, are commonly used—their fur is thought to help wounds heal, their eyeballs are used as a love potion, their meat is said to cure epilepsy, stomach problems or asthma, and their entire body might be soaked in alcohol to make an "energy drink."

Western medicine continues to use thousands of primates in biomedical research laboratories every year. These animals are sacrificed in order to test the safety of drugs intended for human use. The moral dilemmas involved in such experimentation are complex, and critics point out that even closely related species react differently to certain drugs and pathogens. Most laboratory animals are now captive-bred, but this does not resolve the moral issues, and entrepreneurs still try to sell wild-caught primates into this lucrative trade, which has had a significant impact on some populations of certain species. Campaigners urge experimenters on ethical grounds to reduce, refine and replace the use of primates in laboratories, but progress is slow.

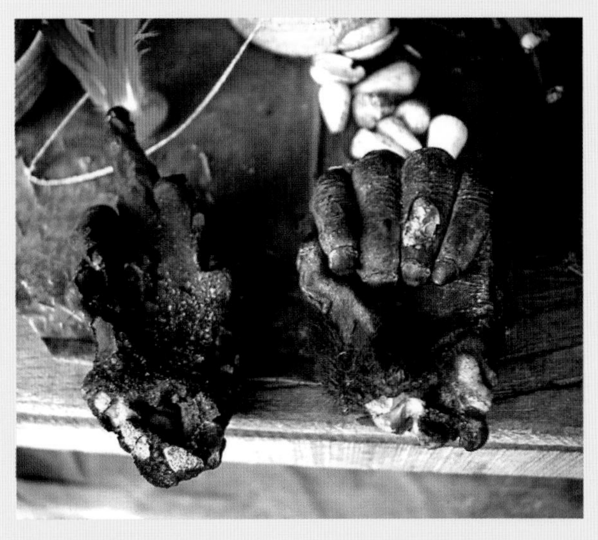

Above In some parts of Congo, people believe that a bit of gorilla finger can cure a sick child, which is why this hand has fingers sawn off.

Facing page Being close to gorilla infants at play is delightful, but may expose the young gorillas to a human virus against which they have no immunity.

Primate Tourism

Tourism is a double-edged sword for conservation. The arrival of tourists can create jobs, bring sustainable finance to preserve an area and educate the visitors, who then return home to spread the word. Equally it can bring disaster, by alienating the local people and changing their culture, transmitting diseases and even destroying the very thing that attracted the tourists in the first place. When primates are the attraction, tourism can be a matter of life or death.

People like to watch primates, and some are prepared to pay to do so. Primates and their habitat need conservation, which costs money. It sounds like a simple win-win situation if the two can be brought together. But most primates inhabit inaccessible places and are wary, if not terrified, of humans. For such a venture to succeed, you need to win the trust of the primates and cater for the needs of the tourists, while at the same time engaging the local human communities so that they also see positive benefits from the development.

Perhaps the best positive example of primate tourism is that of the Mountain Gorilla. The late Dian Fossey's ground-breaking work with Mountain Gorillas during the 1960s and early 1970s was featured in *National Geographic Magazine* and a television documentary, *Search for the Great Apes*. Fossey had developed techniques of habituating several families to the presence of a harmless observer who observed gorilla etiquette, and this enabled her to be accepted by the dominant silverback leaders of each group. Sadly, as our knowledge of Mountain Gorilla behavior and ecology grew, gorilla numbers continued to decline. By the early 1980s, the population in the Virunga Mountains along the northern border of Rwanda, the Democratic Republic of the Congo (DRC) and Uganda was 250 individuals—half what it had been in 1960.

THE POWER OF TV

One of the most popular moments in the history of English-language television occurred when Sir David Attenborough sat in the midst of a family of Mountain Gorillas during the filming of a sequence for the landmark BBC TV series *Life on Earth*. That sequence was filmed in 1978. When Mountain Gorilla tourism began in 1980, there were already large numbers of people keen to visit Rwanda and pay for a permit to "do a David Attenborough." The success of natural history films on television and in movies continues to draw visitors to see wildlife in the wild.

Above An hour spent observing a family of Mountain Gorillas can be a life-changing experience for human tourists, and at this distance, neither party is in danger from the other.

At the end of 1977, poachers speared a famous Mountain Gorilla called Digit, and Fossey launched an appeal for help. She set up the Digit Fund (now the Dian Fossey Gorilla Fund International, based in the United States, and the Gorilla Organization in the United Kingdom). Other conservation groups pitched in and established the Mountain Gorilla Project, which took a three-pronged approach to conserving the gorillas: improve antipoaching measures; educate local communities; and develop gorilla-watching as a tourist attraction. Initially this work was confined to Rwanda, but its success was later emulated in the DRC and Uganda through the International Gorilla Conservation Program. With this level of support, the governments of the three countries were able to reverse the decline in gorilla numbers, develop a successful gorilla tourism industry and change attitudes among communities living in the whole region.

Gorilla success story

What was the key to this extraordinary success? Simply the gorillas themselves, and the creation of the right circumstances for visitors to see them behaving naturally in their own habitat. If anyone before Dian Fossey had suggested that tourists would pay hundreds of dollars to get soaked, sweaty and muddy, not to mention scratched, stung and exhausted, just to spend an hour in the presence of a bunch of hairy apes, they would have been ridiculed. Everyone knew that gorillas were difficult to see and dangerous to approach. But two of Dian's former students—Bill Weber and Amy Vedder—adapted Dian's methods to habituate families of gorillas for tourists to visit on foot, like a researcher. Keeping the gorillas' welfare to the fore, they introduced rules to minimize the disturbance. Visits were limited to one small party per day, for one hour only, and visitors were permitted to get no closer than a sneeze can carry.

Word spread fast and within a year people were lining up to buy permits; within a decade, attitudes in government and local business had been transformed, and the conservation of gorillas began to become a priority. Images of gorillas appeared on the currencies of Rwanda and Zaire (now known as the Democratic Republic of the Congo—DRC) and safari companies began competing for permits to enable their clients to experience an hour of gorilla family life. Even the horrific civil wars, genocide and refugee crises that tore through the region did not halt the gorilla tourism industry for long. As soon as security improved, the tourists returned, and now this multimillion-dollar enterprise plays a central role in the economy of the region. The problems have not completely disappeared, of course, but the risks are being dealt with and, thanks to the dedicated work of courageous local and international conservationists, the population of Mountain Gorillas is slowly but surely increasing year after year.

This constructive approach to primate tourism has now been adapted to several forests with Chimpanzees in Uganda and Rwanda, to Golden Monkeys in the Virunga Mountains, and to Black-and-white Colobus monkeys in Rwanda's Nyungwe Forest. It is also spreading to other countries such as Tanzania and Ivory Coast. The hope is that low-volume, high-value primate tourism can conserve primate species and their habitats while bringing real improvements to the lives of people in neighboring communities.

Some primate-viewing locations have developed out of long-standing associations between local people and the other resident primates. Buddhist and Hindu temples, for example, often have a tradition of feeding monkeys, and some visitors to the temples may be drawn there as much by the chance of seeing the monkeys as to see the monks. However, if these interactions are not well managed, they can result in problems.

In 2007, the publication of a 19-year study of the impact of visitors on Tibetan Macaques in the Mt. Huangshan Scenic Area in China's Anhui province revealed that when tourists fed the monkeys, aggression in the troop increased, leading to more infant deaths. The ranging pattern of a troop is likely to change if artificial feeding becomes a regular event—it must seem like a never-ending fruit tree, and competition for such a resource is intense. If the food provided is mainly domestic crops or cooked meals, the result is probably an increase in crop-raiding and theft from unattended kitchens. And if the tourists themselves begin sharing their packed lunches, every bag will then be seen by alert monkeys as a potential meal, and children especially will be at risk of being bitten.

Health risks

Sharing a meal with a primate of another species can bring two-way health risks. A monkey eating a half-bitten sandwich may catch a cold, flu or digestive-tract infection that could well be deadly and might be transmitted to other members of the troop. Meanwhile, physical contact—particularly a bite resulting from someone trying to hold onto a bag or defend a child—can transmit monkey viruses or bacteria to the tourist's bloodstream. This may turn out to be harmless or it may trigger a major public health crisis, potentially as devastating as HIV, SARS or avian flu.

The evolution of viruses is the focus of many studies, as scientists try to understand how new diseases emerge and old ones develop resistance to drugs. In another study of primate-human interactions, researchers concluded that out of every thousand people who visit a monkey temple in Bali, Indonesia, six are likely to be bitten by a monkey and infected with simian foamy virus (SFV). This virus has not yet caused a disease in humans, but viruses mutate and it may just be a matter of time. Moreover, the problem is becoming worse as such religious sites become overcrowded with populations of "refugee" macaques fleeing from deforested areas. The mixing of primate populations as a result of humans destroying their habitats also leads to mixing of different strains of viruses, so new combinations may arise with unknown clinical effects in humans.

Obeying the rules

Some macaque troops in Japan are visited by many thousands of people every year and, in contrast to experiences elsewhere in the world, both monkeys and humans seem to get along fine. At Arashiyama, for instance, on a hilltop overlooking the ancient city of Kyoto, macaques are fed grain scattered around the tourists. The monkeys concentrate on filling their cheek-pouches with grain and the tourists concentrate on taking photographs, with neither disturbing the other. Perhaps due to the Japanese culture

Right This group of Japanese Macaques is busily gathering grain scattered on the ground for its benefit at Arashiyama, high above the city of Kyoto, Japan, unconcerned about the presence of human visitors.

of respecting rules, no one ever feeds the monkeys outside, so no monkey tries to steal people's bags. If a visitor does want to feed a macaque, there is a hut with wire mesh over the windows and special food inside; the macaques know that if they hang on the outside and reach in, food will be given to them. Once again, both species seem satisfied with the arrangement and appear to understand and follow the "rules." The lesson here is that interactions between human and nonhuman primates can be beneficial to both, if carefully managed.

PRIMATE-WATCHING GUIDELINES

RESPECT THEIR HABITAT
Large numbers of visitors may have a negative impact on delicate ecosystems, so it is important to heed the motto "take only photographs and leave only footprints."

DON'T GET TOO CLOSE
Just like humans, other primates have a sense of their own personal space. Gauge how they react as you approach, and be prepared to stop when they start to look nervous: usually this will be a safe distance for both parties.

DON'T PURSUE THEM
If primates move away from you while you are watching them, wait a while before approaching again, otherwise they will get the impression you are chasing them.

DO USE A GUIDE
If there are official guides, or local people who can help guide you, hiring them improves your chances of seeing the primates and encourages the local community to value their primate neighbors.

DON'T FEED PRIMATES
Never offer human food or eat in front of wild primates. The next person carrying a bag like yours might be attacked because one of the troop learned that it might contain food. If there is an organized system of feeding appropriate foods, you can judge whether you want to participate, but such places can have a negative impact on primate health and behavior if not well managed.

IT'S RUDE TO STARE
As in human society, a long, uninterrupted stare is likely to be found intimidating and may be mistaken for a challenge. If you catch a primate's eye, look away and glance back.

COUGHS AND SNEEZES CAN KILL
If you visit wild primates while you are sick—even with only a runny nose or diarrhea—you risk killing them. They are close relatives, and therefore many diseases are easily transmitted between us and other primate species.

DON'T TRY TO PET THEM
You may think you are being friendly, but a primate is likely to interpret any attempt to touch it as aggression. It may try to bite you, but whether it does or not, such close contact risks exposing both you and the primate to each other's viruses, bacteria and even parasites.

A FINAL WARNING
Don't stand directly beneath arboreal primates—if you do, you are effectively standing at the bottom of a long-drop toilet!

Primate Conservation

Primates are in trouble. Ironically, as our knowledge of them increases and our respect for their intelligence grows, their numbers dwindle. Not every species is declining, but the 2007 Red List of Threatened and Endangered Species includes 114 primate species. Of these, 21 are critically endangered, 47 are endangered and 46 are considered vulnerable to extinction. Unless humans mend their destructive ways over the next few decades, the primate family tree will be severely pruned, and even the human branch may ultimately be threatened.

The Hainan Gibbon is probably the most endangered primate in the world. Out of 20 individuals known to survive in Bawangling National Nature Reserve in southern China's Hainan Province, only three are breeding females, although conservationists were hopeful in 2007 that a young male and female that had been spending time together might form a new family group—the first for years. Efforts to protect the gibbons and improve their habitat have shown some results, because this tiny population is actually an increase on the 15 individuals counted in 2001. Nevertheless, this contrasts with the estimated total population of 2,000 of these gibbons in the 1950s.

Twin dangers

Deforestation and hunting have been the two major threats to Hainan Gibbons, indeed to most primates and other forest wildlife. Conservationists tackle these dangers by passing and enforcing laws to protect the species and their habitats, and by educating people whose daily decisions are causing the problem and finding them alternative livelihoods or food. It is extraordinary that about 50 percent of the world's species live in tropical rainforests and campaigns to save the rainforests have been running for decades, yet despite this, deforestation continues apace. Why? Because the economic pressure driving

Facing page Beautiful but critically endangered, this Hainan Gibbon in southern China has a future only if hunting and forest destruction is halted.

deforestation is measured in billions of dollars, a thousand-fold greater than the extent of financial resources being devoted to funding conservation.

With new satellite images available on the Internet, it is even possible to watch deforestation online. This does present new opportunities for law enforcement, and members of the public can now play a role in drawing attention to illegal logging in, for example, national parks. Even where a forest is still standing outside protected areas, selective logging results in degradation of the habitat, and logging roads provide easy access to commercial hunters. These professionals hunt not only to supply logging camps: they also use the logging trucks to carry tons of bushmeat to markets and restaurants in distant cities.

The bushmeat trade

Many rural people move to cities to find work; those who succeed then have money to buy foods they remember from their youth—fruits and other foods harvested from the wild, including bushmeat. In some parts of Africa and Asia, people regard primates as just another wild animal to shoot for the pot. And because they spend much time up trees, primates often make a clear target for a hunter with a shotgun or a traditional blowpipe.

Above This frightened, confused Chimpanzee infant is illegally being offered for sale in the bushmeat market where the meat of its newly slaughtered parents is also illegally on sale, in Libreville, Gabon, Central Africa. Infants are worth more alive than dead, but many die before they can be rescued.

Below Smoked monkey meat is normal fare in forested parts of Africa, but commercial hunters with modern equipment are emptying the forests of wildlife.

Below Logging, seen here in northern Borneo, poses three threats to forest-dwelling primates: destruction of their home; the introduction into the forest of a workforce who may eat bushmeat; and the provision of transport for bushmeat and live primate infants on logging trucks and trains to distant urban markets.

CLIMATE CHANGE
A NEW CONSERVATION OPPORTUNITY?

The logic is simple: to save the primates, we must save their habitats. Primate habitats mostly comprise the planet's three "green lungs"—the tropical forests of Central and South America, Africa and Southeast Asia. As the world wakes up to the potential consequences of climate change, fear of an immense global catastrophe has brought the ecological services that these three forested areas provide to the fore. At last there is a growing sense of urgency in efforts to slow, and ultimately reverse, deforestation.

Reading some conservation literature, one might conclude that a species' beauty and potential for giving humans pleasure was the main reason for conserving it. But monkeys and apes are not just important because they are cute and intelligent social mammals. Primates also perform a service to the planet. They, together with elephants, parrots and other fruit-eating animals, are keystone species in their habitats, principally because they disperse the seeds of the next generation of trees in their droppings. The trees we fell today for our garden furniture and hardwood paneling were effectively planted by animals many centuries ago. This means that to save tropical forests in the long term, we need to dedicate resources to the protection of the "gardeners of the forest"—the apes, elephants and other seed-dispersal agents (*see* pages 18–19).

This is not, of course, just about saving charismatic megavertebrates—large animals that most people find cute or beautiful. About 50 percent of all known species live in tropical forests or, more correctly, play a part in the ecology of tropical forests. These forests also perform a pivotal role in storing carbon and carbon sequestration (the processes that remove carbon dioxide from the atmosphere). Moreover, losing the forests doesn't just mean losing their role as climate regulators. Forest destruction and degradation account for nearly a fifth of global greenhouse gas emissions—much more than that contributed by the transportation sector.

Cutting down forests is a double loss because the process adds to the very problems we need the forests to help solve. Forests store carbon not just in the wood of the trees, but also in the soil—especially in tropical forests growing on peat swamps. Such forests release many centuries of stored carbon if they dry out through a change in patterns of rainfall. Many people hope that if this carbon can be traded as carbon credits in the carbon markets of the world, the money raised could pay for conservation, which would not only save endangered species but also keep the forests healthy so that they can continue absorbing carbon dioxide and regulating the climate.

Facing page Mount Visoke's rain-fed crater lake lies at an altitude of about 12,000 ft. (3,700 m). Visoke is one of a chain of volcanoes that form the Virunga Mountains along Rwanda's northern border with Uganda and the DRC. The lushly forested slopes are home to Mountain Gorillas and Golden Guenons; these species play a vital role in maintaining the health of the forest and, by extension, its ability to store carbon and maintain the water cycle for the surrounding farmland.

If the killed primate is a mother carrying an infant, the hunter may take the baby as a pet or to sell into the live animal trade (which is illegal in almost every country except in special circumstances).

Despite being legally protected, primates continue to be hunted apace in many countries. Legal protection takes different forms: primates may be included in the official list of species not allowed to be hunted; or their habitat may have been made into a national park or reserve. Unfortunately, many developing countries do not have the resources to enforce their wildlife laws, so poachers and illegal traders are seldom arrested and brought to court.

Viral threats

Hunting, logging and the conversion of forest to agriculture directly harm primates, but these practices may also lead to increased threats from viruses. Forest degradation, for example, presents increased opportunities for viruses to jump species as neighboring troops of primates are driven into each other's territories, immunity is reduced by rising stress levels and injuries from fighting potentially lead to cross-species infections. In turn, these infections potentially lead to new strains of viruses, which may find a way into the human population. It is thought that HIV made such a jump from Chimpanzees to humans in Central Africa, and Ebola outbreaks are thought to have similar origins. Unlike simian immunodeficiency virus (SIV), which does not cause clinical symptoms in Chimpanzees, the Ebola virus kills up to 90 percent of hominids it infects—humans, gorillas and Chimpanzees. The 2007 listing of Western Lowland Gorillas as critically endangered was due to an estimated 60 percent decline in their numbers over 20 years due to Ebola infection and bushmeat hunting. Today's primate hunters are also more likely to transport any viruses, in bushmeat or caught while butchering, out of the forest and into urban areas, where the risk of an epidemic becomes much higher.

Climate change

The most significant of the emerging threats to primates is climate change. If sea levels rise, coastal forests may be lost, along with any endemic species living there. If rainfall patterns change, the particular local conditions that enable forests to grow may be disrupted in a relatively short period of time, defined as anything less than the time it takes for new trees to grow at the more favorable edge of the changing habitat. Species living in forests that have been segmented by human development will be unable to adapt as their habitats dry out and will perish.

On the other hand, there are hopes that carbon finance may bring new revenues and jobs to the forestry sector, in an attempt to conserve forests to mitigate climate change. This is why some conservationists think that saving the primates—and their habitats—could save the world (see box). The alternative scenario—"business as usual"—could be catastrophic for us all.

Facing page Human-gorilla friendships, as seen here in the John Aspinall Foundation rehabilitation project in Congo, can be rewarding. When such gorillas return to the wild, steps are taken to avoid encounters with terrified villagers.

In the Field

Primate behavior is fascinating, and watching primates as a pastime or career is open to anyone in any country. The easiest primate to observe is, of course, the human primate. Anthropology is the study of human behavior and societies, but many anthropologists go on to study nonhuman primates, just as some primatologists apply their zoological approach to study humans. Such comparative studies can help us to understand our own behavior better, as well as our relationship to other species and the natural world.

Scientific observations are supposed to be objective and rational and to test theories to explain the phenomenon being observed. But all scientists are human, so inevitably are a product of their own culture, experience and education. This means that great efforts have to be made to avoid anthropomorphism—that is, interpreting animal behavior in human terms—especially value judgments based on human morals or culture. Indeed, for much of the 20th century most scientists were so afraid of being accused of anthropomorphism that they failed to recognize many of the behavioral patterns that human and nonhuman primates share.

Yet one of the prime motivations of the Kenyan anthropologist Louis Leakey (1903–1972) in initiating studies of great apes was to search for behaviors we had in common with them. He spent his life digging up the fossil bones and tools of human ancestors and wanted to know how the hominids he was studying behaved. He reasoned that behaviors that are essentially the same in apes and humans are most likely to have been inherited from our common ancestor, whereas differences must have evolved to cope with the different conditions that produced the species we see today.

Above Savanna species are easier to study than forest species, but you still need patience to win their trust. Here, Professor Lynne Isbell observes a Patas Monkey in Kenya. It took six months of following a troop before she could approach closely.

Old and new skills

The first close observers of wild animals were mostly hunters, and even today much of the knowledge of primates that have not yet been the subject of scientific study is in the minds of the people who share their forest and watch them. Ironically, the modern scientist armed with binoculars, laptop computer and GPS device is often dependent on the old-fashioned field skills of an illiterate forest-dwelling hunter to locate the animal to be studied. When these two very different skill sets are brought together, however, the results can transform our understanding of the species being studied. Moreover, by hiring a hunter and using his skills for nonlethal research or tourism, his motivation for killing the animals in question changes. In many cases, conservation success has followed in the wake of such career changes; whole communities can benefit from the employment and educational opportunities that a long-term field research center can bring.

The results of such research can, for example, be used to inform decision makers who determine land use. If the primate species or behavior being studied is the subject of popular books, magazine articles and documentaries, it can attract more attention to the site. This, if carefully managed, can provide the basis of further research or ecotourism projects. As long as the resulting development and disturbance does not impact negatively on the species concerned, significant benefits can flow from what began as a simple research project. More and more primate species are being studied and habituated for tourism, and there are growing numbers of primate watchers, who, like birdwatchers, want to seek out as many kinds of primates as they can. The challenge is to harness this interest to help conserve the primates being watched—and to improve the lives of the people living in and around their habitats.

Prosimians

Main image Typical lemurs, such as this Red-fronted Brown Lemur (*Eulemur fulvus rufus*), a subspecies of the Brown Lemur, resemble adapids—ancestral primates from 40 million years ago.

Introduction: Prosimians

Take a walk at night in a forest or woodland in Africa, India or Southeast Asia and shine a flashlight around the trees and bushes. With luck, you will catch sight of a pair of glowing round eyes reflecting the light back at you—a nocturnal primate. Take a walk at any time of day or night in Madagascar, and the only nonhuman primates you'll see are lemurs. In both cases, the animals you are watching are prosimians—the living primates that most resemble the earliest fossils of primates from the Eocene epoch, between 35 and 55 million years ago.

On the primate family tree (see pages 28–29), the prosimian branches stem back to the lowest, and therefore the oldest, part of the trunk (pro-simian means "before apes"). Today, we see three bushy ends to this branch and one long separate "twig." The bushy ends represent the lemurs, the galagos, or bushbabies, and the lorises, all of which belong to the suborder Strepsirrhini. These are the wet-nosed primates, which have a longer snout than most other primates and a larger olfactory lobe in the brain, giving them a very well-developed sense of smell. They use scent-marking as well as visual displays to communicate. Anyone who thinks that bushbabies would make cute pets is in for a shock—they urinate on their hands and feet and use them as scent markers, so these creatures would soon wreak havoc as they climbed all over their owners and the furniture. In the wild, however, this behavior keeps them in touch with the movements of their family, friends and rivals. Studies of Pygmy Slow Lorises and galagos in captivity show that they can recognize specific individuals by their scent.

Physical features

Strepsirrhine primates share a number of features that differentiate them from other primates. For example, they have a grooming claw instead of a flat nail on the second toe of each foot, and, in every species apart from the Aye-aye, the lower canine and incisor teeth are clustered together to form a tooth-comb, which is also used for grooming. The Aye-aye is a special case—it is so different from other lemurs that it is classified in its own family, the Daubentoniidae, and its strange appearance has given rise to many superstitions and beliefs among the people of Madagascar.

Somewhere between the prosimian and anthropoid branches of the primate tree lies the aforementioned long, thin twig—the tarsiers. The exact relationship of the tarsiers to other primates is still being debated. Although they are nocturnal insect-eaters that cling to vertical stems and leap, rather like bushbabies, they possess some important physical differences that suggest a separate evolutionary origin. In particular, they lack the wet nose of other prosimians, so taxonomists now include them with monkeys and apes in the suborder Haplorrhini (the dry-nosed primates). Tarsiers also lack the special reflective layer in the back of the eye, known as the *tapetum lucidum*, that improves nocturnal vision in

Left Their large eyes, large ears and a wet nose enable nocturnal prosimians—such as this Senegal Lesser Bushbaby, which is poised before jumping—to forage for food and locate a mate in the night.

bushbabies, lorises and lemurs. (The *tapetum lucidum* produces the mirrorlike eyes that reflect a flashlight beam at night; this same structure is also found in certain carnivores, such as cats.)

Ecological niches

Prosimians appear to have found a number of ecological niches and—except in Madagascar—have stuck to them with only minor variations. Fossilized primate remains from China and North Africa that date back 40 million years form two groups: the omomyids, which look very much like today's nocturnal prosimians, especially the tarsiers; and the slightly larger adapids, which look like lemurs. The recent discovery of lorislike fossils in Egypt, dating back 40 million years, suggest that the branches leading to today's prosimians diverged from each other at this early stage and have not changed very much since that time. In Madagascar, however, where there are no monkeys or apes to compete with for food and territory, prosimians evolved into many new forms to occupy the daytime, fruit-, leaf- and insect-eating niches. This makes Madagascan primates of huge interest to evolutionary biologists. The island is often described as a giant evolutionary experiment that illustrates how natural selection can act to produce a great diversity of species.

Various different lemurs fill the ecological roles of woodpeckers, rabbits, pandas and raccoons in Madagascar, as well as playing the equivalent roles of monkeys and apes. Until relatively recently, there were even Madagascan lemurs the size of gorillas—now known only from subfossil (partly fossilized) skeletal remains. Their extinction—together with around a third of the known lemur species—occurred during the first 1,000 years of the 2,000-year period since the arrival of humans in Madagascar.

Today, urgent efforts are being made to slow the clearing of forests for agriculture and stop the hunting of lemurs, in the hope that the species that survived the early onslaught by humans can survive in the future. Hanging over all such conservation efforts, however, is the dire threat of climate change. If rainfall patterns change, the distribution of suitable forest habitats will change too, creating new challenges for conservationists.

Below Bounding bipedally, this Verreaux's Sifaka provides a wild ride for her infant, whose knowledge of their habitat is gained during such searches for food.

Lorises and the Potto

Resembling goggle-eyed, animated fluffy toys, lorises and Pottos are among the least studied of all primates. They have long been characterized as "nocturnal and creepers" without much of a social life, but in recent years new studies have started to throw more light on these secretive animals, and many previous assumptions have been proved wrong.

Left This Greater Slow Loris has perfect poise as it balances on twigs, stalking prey that moves slowly enough for it to capture.

Below Long, gangly limbs characterize the endangered, delicate-looking Red Slender Loris from southwestern Sri Lanka.

In terms of their locomotion, lorislike primates can be divided into two main groups. The "leapers" are the African bushbabies (see pages 54–57). The "creepers" include the stick-limbed slender lorises of India, the slow lorises of Southeast Asia, and the angwantibos and Pottos of Africa. Members of the latter group clamber along and between branches with slow, deliberate movements, hunting for slow-moving invertebrates, many of which are poisonous, as well as birds' eggs and fruit.

The false impression that lorises and Pottos are mostly solitary is due largely to the difficulty in observing their behavior at night in dense tropical forest, without disturbing them. Moreover, many of their social interactions involve the senses of hearing and smell, and some are beyond the ability of human senses to detect. Lorises and Pottos were once thought to vocalize much less than bushbabies, but it was subsequently discovered that some of their calls include ultrasonic sounds, which can be registered by the human ear only with the help of a bat detector (a device that reduces the pitch of an ultrasonic sound until it is audible to humans).

Many of the 34 species of lorises and Pottos appear very similar to the human eye and are thus known as cryptic species. It is only with detailed DNA studies that the complexity of their taxonomy is being revealed, so it seems likely that more new species will be described. Early books would differentiate only between the Slender Loris (*Loris tardigradus*) of India and Sri Lanka and the Slow Loris (*Nycticebus coucang*) of Southeast Asia, for example. The Slow Loris was considered to be widespread from northeast India south through Malaysia to Indonesia and the southern Philippines, and regional varieties were given the status of subspecies rather than full species. But in 1994, on genetic grounds, the orange-brown lorises found in Cambodia, Laos, southern China and Vietnam were recognized as a separate species: the Pygmy Slow Loris (*Nycticebus pygmaeus*). Another species, *N. intermedius*, was described from the same area, but its distinguishing characteristics turned out simply to be seasonal changes in the coat and body weight of the Pygmy Slow Loris.

New species

In 2003, further genetic research on several populations considered to be subspecies of the Greater Slow Loris (*Nycticebus coucang*) confirmed what some had long suspected, and three new loris species were described. Their most obvious distinguishing characteristic is the pattern—or lack of it—on top of the head and neck, between the dark eye-rings and the dorsal stripe. The Bornean Slow Loris (*N. menagensis*) has a soft brown crown and lives in Borneo and the southern Philippines; the Bengal Slow Loris (*N. bengalensis*) has a pale crown and is found from India to Vietnam; while the Javan Slow Loris (*N. javanicus*) has striking, badgerlike stripes connecting the eye-ring to a thick black dorsal stripe.

All loris species are threatened by the destruction and degradation of their forest habitats, and, although trade in them is illegal, they are killed for their meat and supposed medicinal properties. They are also captured for sale as pets. The appeal of a loris as a pet is easy to understand at first glance, but they do not fare well in captivity and most die within a short time of capture. If they do survive to be bought by a doting owner, there is an unseen danger. Slow lorises lick the insides of their arms, where glands secrete a sweatlike substance that contains toxins to paralyze their prey. The saliva activates these toxins, so a bite from a loris can be doubly dangerous. Not only do lorises have long, razor-sharp canine teeth that inflict deep lacerations, but some people react to loris saliva by going into anaphylactic shock, which, if untreated, may cause death. In 2007, all international trade in slow lorises was banned.

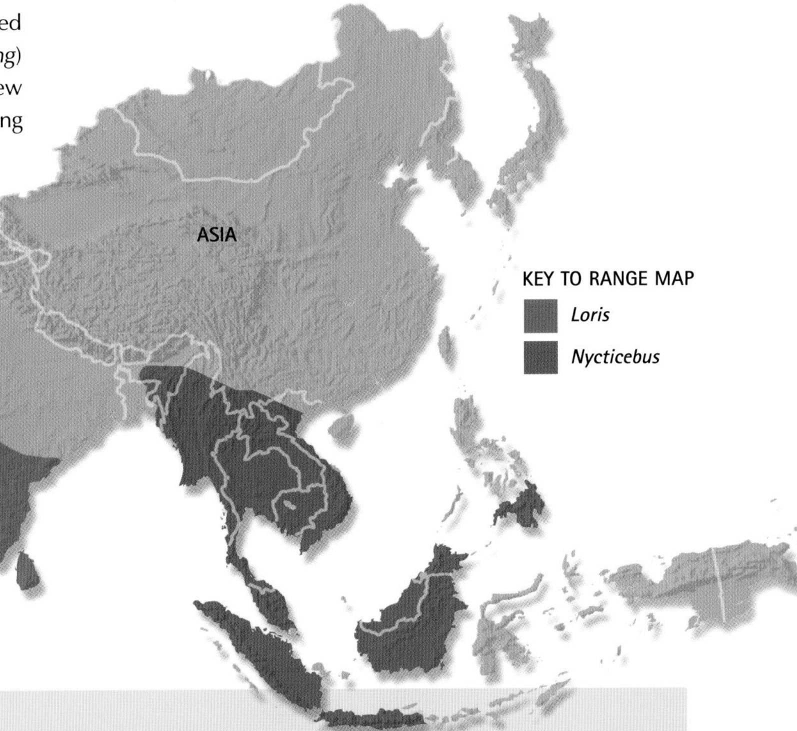

LORIS SPECIES

Scientific name	Common name	Where they live	Red List	CITES
Loris lydekkerianus	Gray Slender Loris	Southern India✪, Sri Lanka✪	NT	II
Loris tardigradus	Red Slender Loris	Southern India, Sri Lanka✪	EN	II
Nycticebus bengalensis	Bengal Slow Loris	Bangladesh, Burma, Cambodia, China, India, Laos, Thailand✪, Vietnam✪	DD	II
Nycticebus coucang	Greater Slow Loris	Sumatra (Indonesia)✪, Peninsular Malaysia✪, Thailand✪	LC	II
Nycticebus javanicus	Javan Slow Loris	Java (Indonesia)✪	DD	II
Nycticebus menagensis	Bornean Slow Loris	Borneo✪	NE	II
Nycticebus pygmaeus	Pygmy Slow Loris	Cambodia, China✪, Laos	VU	II

RED LIST: CR = Critically Endangered EN = Endangered VU = Vulnerable NT = Near Threatened LC = Least Concern DD = Data Deficient NE = Not Evaluated ✪ = Best place to watch

PRIMATE PROFILE
LORISES, POTTO AND ANGWANTIBOS

SIZE Head and body length: 6½–15½ in. (17–39 cm); tail: 14½–40 in. (37–100 cm); weight: 3½ oz.–3½ lb. (100–1,600 g)

APPEARANCE Coat color variable (browns and grays); short, soft fur, short tail

HABITAT Primary and secondary forest; plantations and wooded savanna

DIET Fruit, seeds, leaves, bark, fungi, gums, shoots, flowers, birds' eggs, insects, small vertebrates, invertebrates such as snails and crayfish

LIFE HISTORY Gestation: about 6 months; sexual maturity: male not known, female 3.5 years; life span: up to 33 years

BEHAVIOR Solitary, nocturnal, slow foraging in trees

Facing page A Potto emerges at dusk from its resting hole in a hollow tree, to forage in the forest near Epulu, Democratic Republic of Congo.

Speed and prey

Lorises, Pottos and angwantibos have a powerful grip and can cling to branches for hours without tiring. Their index fingers are small—in Pottos they are barely present—and their thumbs are positioned opposite the other fingers. Their wrist and ankle joints are more flexible than those of other primates, allowing them to clamber more effectively between branches and vines. They have very short tails and their arms and legs are of similar lengths—in contrast to bushbabies and tarsiers, which have longer legs for jumping and long tails to balance while leaping and landing.

Being slow-moving does limit the kind of prey these primates can catch. Instead of taking the moths, beetles and crickets favored by bushbabies, the lorises, Pottos and angwantibos focus on less mobile but foul-tasting prey, apparently immune to the chemical defenses of their victims. For example, they are not deterred by ants squirting formic acid, millipedes that exude a noxious liquid when attacked or caterpillars with irritant hairs. Angwantibos specialize in eating such caterpillars, but before consuming each one they hold it by the head and wipe their hands down the body carefully, removing the hairs. As well as invertebrates, Pottos will catch birds and bats. Pottos and slow lorises also feed on fruit and gum, and one slow loris population was found to depend on nectar from palm tree flowers for nearly one-third of its diet. Slender lorises, in contrast, are among the most predatory primates, with up to 100 percent of their food being animal prey. Interestingly, young Pottos and angwantibos learn what to eat by snatching food from their mother and examining it, cocking their head in a characteristic way.

Defense

When attacked, these sluggish primates sometimes deliberately let go of the branch and fall, but Pottos have an additional and unique defense—apophyseal spines. Several of their neck vertebrae (numbers 3 to 9) have long, bony parts that actually protrude from the skin as spines. When threatened, a Potto tucks in its head to present the spines, and above them its shoulder blades form a bony shield. Lorises, Pottos and angwantibos are usually alone when encountered, but behavioral studies are beginning to reveal a more complex social life than previously suspected. Adults have been seen spending varying amounts of time with other adults, and male slender lorises often play with infants from neighboring sleeping groups if they encounter them. Male territories are larger than those of females and include more than one female territory. Females announce their receptivity by urine-marking and, in some species, by giving "estrus calls"—in slow lorises, these are frequent, high-pitched whistles.

Pottos and angwantibos give birth to a single infant, lorises to one or two. There are sometimes two births in a year. Instead of carrying the infant when foraging, angwantibo and loris mothers "park" their offspring on a branch or in a tree hole, although Pottos rarely do so. After foraging all night, lorises, Pottos and angwantibos sleep in tangled thickets or forked branches, sometimes in male-female pairs with offspring; in some species, the adults sleep alone unless a mother has an infant with her. Sleeping during the day makes these primates particularly vulnerable to forest clearance. Primates that are awake during the day may have a chance to flee from forest fires or chainsaws, but nocturnal primates are likely to be destroyed along with their habitats.

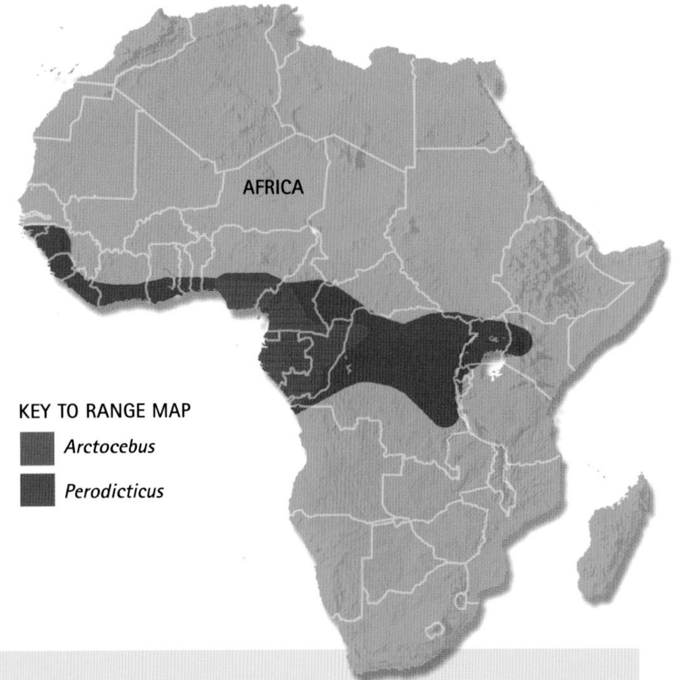

ANGWANTIBO AND POTTO SPECIES

Scientific name	Common name	Where they live	Red List	CITES
Arctocebus aureus	Golden Angwantibo	Gabon✪	NT	II
Arctocebus calabarensis	Calabar Angwantibo	Nigeria✪ to Congo✪, Gabon	NT	II
Perodicticus potto	Potto	West Africa to DRC✪, Uganda, Kenya✪	LC	II

RED LIST: CR = Critically Endangered EN = Endangered VU = Vulnerable NT = Near Threatened LC = Least Concern DD = Data Deficient NE = Not Evaluated ✪ = Best place to watch

PRIMATE PROFILE
BUSHBABIES

SIZE Head and body length: 4–18½ in. (10.5–47 cm); weight: 7 oz.–3¼ lb. (200–1,500 g)

APPEARANCE Large eyes and ears; long, bushy tail; slender body; coat color variable

HABITAT Wide range of forest and woodland habitats; scrub and savanna

DIET Fruit, gums, leaves, buds, seeds, small vertebrates, insects

LIFE HISTORY Gestation: about 4 months; sexual maturity: 18–24 months; life span: up to 15 years

BEHAVIOR Tree-living, nocturnal, solitary and group foraging; some species have complex social structures

acing page Like all bushbabies, Demidoff's Bushbaby anoints its
ands and feet with urine, leaving a scent trail wherever it climbs.

Bushbabies

he African night is a symphony of many animal calls—frogs plink, cicadas buzz, and hyenas whoop and laugh—but there is one call that is unnervingly like that of a human infant. This is he call of a bushbaby, or galago. These active, nocturnal primates are found in forests and woodland across most of sub-Saharan Africa. Although some of the 15 known species look alike, each one has a distinctive "loud call" that alerts neighbors to the caller's whereabouts, and this can be used to identify the species.

Bushbabies are amazingly agile prosimians—they can easily cling onto vertical as well as horizontal supports and can use their powerful back legs to leap across gaps in the canopy up to 20 ft. (6 m) wide, using their long bushy tail as an aerodynamic stabilizer. This form of locomotion is known as "vertical clinging and leaping" and is more commonly used by the smaller bushbabies. Larger species can also jump, but they seem to prefer running along branches on all fours like monkeys.

Specialist diets

Bushbabies eat all kinds of foods, although the larger species tend to take more fruit and the smaller species catch more insects. Prey is detected using their large, mobile ears and good nocturnal vision, and is caught with the hands. Every species, however, eats some tree gum scraped from bark using the tooth-comb (clustered front teeth in the lower jaw). The sticky gum is digested in the cecum (a pouch at the beginning of the large intestine), which is elongated in bushbabies and contains microorganisms able to break down the complex sugars in the gum. It is this ability that enables bushbabies to survive the long winters in temperate South Africa, when fruit, flowers and insects are scarce. One new species, recently discovered in southeastern Tanzania, drinks nectar from flowers, and may act as a pollinator in its woodland habitat.

Due to differences in diet and behaviour between species, in some habitats a number of bushbaby species live alongside each other. For example, in Central Africa's primary rainforests, from Cameroon to Congo, Allen's Bushbaby has a diet of about three-quarters fruit to one-quarter invertebrates and frogs. It lives in the undergrowth, preferring a height of 3–6 ft. (1–2 m) above the forest floor. Above this species, among the tangled thickets at a height of around 35 ft. (10 m), one can glimpse the smaller, more agile Demidoff's Bushbaby. The Demidoff's diet includes almost the reverse proportion of animal and plant matter, with about 10 percent being tree gum and 20 percent fruit. In the same forests, needle-clawed bushbabies are the real gum specialists—tree gum

Above A Thick-tailed Greater Bushbaby eats fruit at a forest feeding station in Tanzania. Such feeding stations offer visitors a reliable opportunity to observe and photograph these elusive, nocturnal primates.

5 percent fruit. These species' clawlike, keeled fingernails enable them to cling to broad tree trunks, and their enlarged canine and premolar teeth are used to scrape for gum, reaching the bark that other bushbabies cannot reach.

There were until recently thought to be six species of bushbaby in one genus, *Galago*. When researchers began to study bushbabies in the wild, however, they discovered that there were many more species—over 20 at the latest count, divided into five genera. More species are expected to be described as research continues.

The smallest of the bushbabies, the Rondo Bushbaby from the lowland forests of southeastern Tanzania, is one of several new bushbaby species that were described in 1997. It has a long reddish tail and long, dark ears. Its belly is pale yellow, deeper yellow under the chin, and its face has a white stripe between dark eye-rings that are less obvious than in its close relatives. But it was the Rondo Bushbaby's distinctive call that led to it being

Below Multiple images show how the bushbaby's hind limbs provide power for the jump, then come forward to act as shock absorbers as the animal lands.

Social leapers

Organizing a social life in the dark can be challenging, which is why, perhaps, so many species of bushbabies look alike. They don't tell each other apart by sight, but by smell and sound. Field studies of such behavior are difficult for visually oriented humans, but new research techniques have confirmed that bushbabies are neither solitary nor simple. They live in a social matrix as complex as those of many diurnal primates.

How can one study small, virtually identical, fast-moving primates in dense forests in the dark? Before the advent of night-vision binoculars, infrared cameras, lightweight radio-tracking devices and DNA analysis, most research into bushbabies was either carried out on captive specimens or involved sound recordings in the dark. Despite the efforts of pioneer scientists, the results were, unsurprisingly, incomplete. In recent years, however, an increasing number of research sites using new technology have led to a reappraisal of bushbaby society.

In most bushbaby species, males have large home ranges overlapping those of several related females, which occupy smaller home ranges. During the day, the females (some with young) may sleep together in a cluster, but they act aggressively toward females from neighboring, unrelated sleeping clusters. Males are also aggressive toward rivals, but may tolerate smaller, low-ranking males. This suggests that an alpha male might be the one that fathers the offspring of all the females in his territory. However, new paternity studies of the Southern Lesser Bushbaby, using DNA fingerprinting, have shown that the alpha male did indeed father the majority of infants, but not all. And some females gave birth to twins fathered by different males. Sleeping arrangements also vary; in some larger species, such as the Thick-tailed Greater Bushbaby, the males join three or four females and offspring in the sleeping cluster in a nest in tangled vegetation. Smaller bushbabies seem to prefer tree holes or a forked branch in a tangled thicket. The species that gathers in the greatest numbers

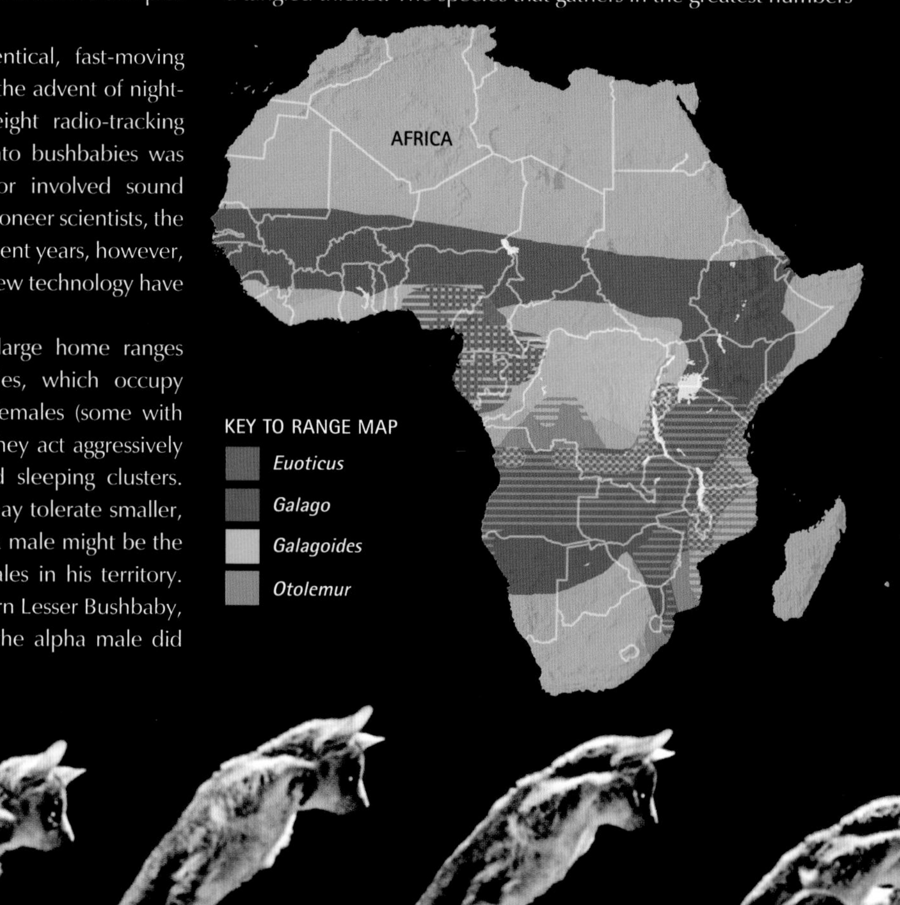

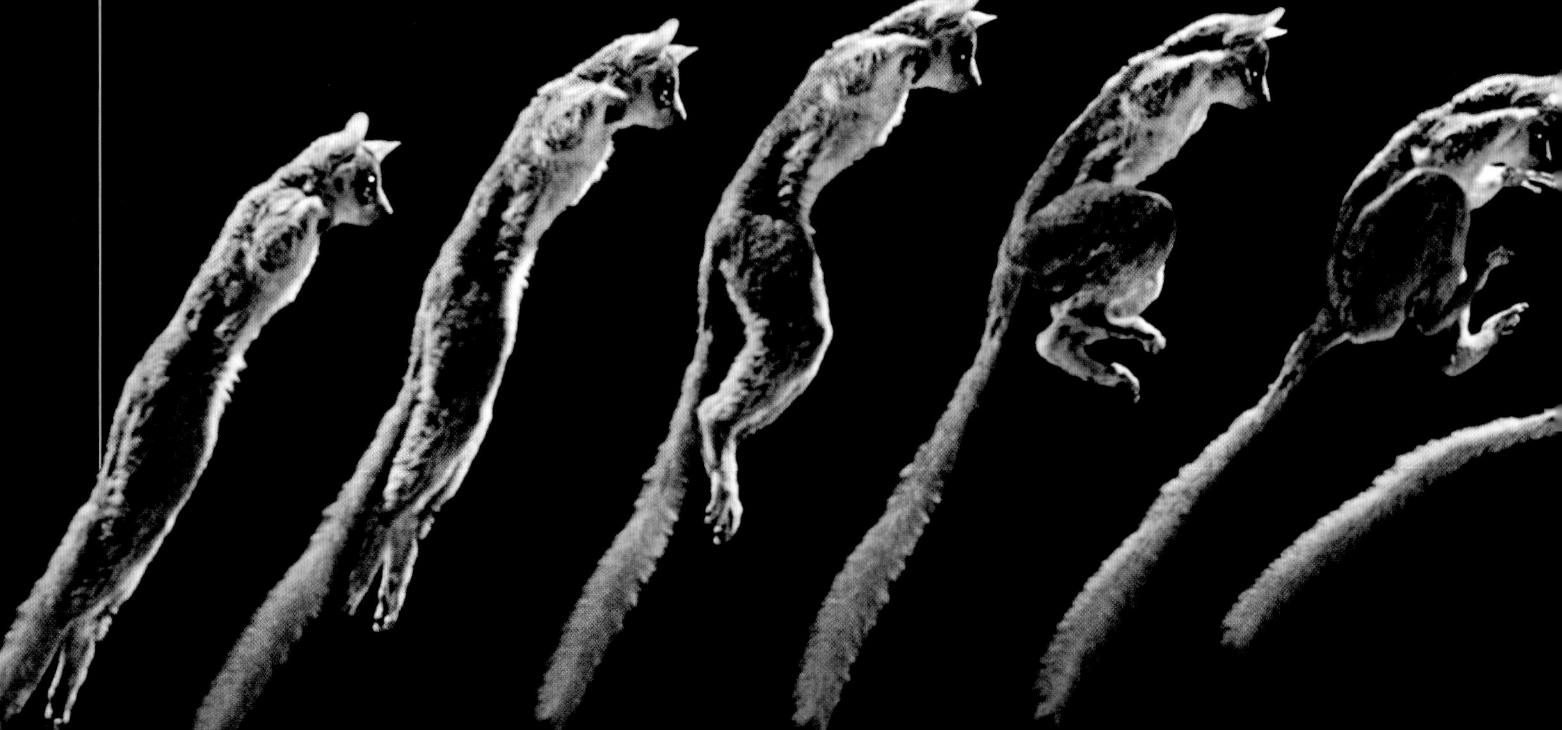

t its sleeping sites is Demidoff's Bushbaby, with up to 10 females lus offspring recorded in a single spherical leaf nest, although in his species the male often sleeps alone.

dentification

Aany bushbaby species, especially males, urinate on their hands nd rub the urine on their feet so that they leave a scent trail. Bushbabies also appear to be able to recognize individuals by their alls. Thus, when two bushbabies meet in the dark, the chances re that they will already know who is approaching. Nevertheless, hey will sniff each other's noses, then faces, as if to make sure r to reinforce their knowledge of the other. Some species spend p to 50 percent of their time in the vicinity of another adult—oraging, feeding or interacting. Most grooming and mutual licking takes place inside the nest. Vocalization also helps maintain th social matrix, and this is true over long distances as well as in close range interactions. Up to 25 separate calls have been recorded i bushbabies. Some types of calls are similar in all species, others ar more distinct and vary between species. Sharp, loud cries are use to raise the alarm if danger is detected, and bushbabies sometime work together to mob a predator by calling loudly in unison. Loud rolling advertising calls that carry far are used either to attract mate or deter a rival—these are very useful for human observers Just as a birdwatcher learns birdsong to identify species of bird, s an experienced primate watcher can identify primate species b their calls, even if the animal is out of sight. As the remote forest of Africa are opened up for scientific research, more species o bushbabies will doubtless be described.

MAIN BUSHBABY SPECIES

Scientific name	Common name	Where they live	Red List	CITES
Euoticus elegantulus	Southern Needle-clawed Bushbaby	Cameroon, Gabon✪	NT	II
Euoticus pallidus	Northern Needle-clawed Bushbaby	Cameroon✪, Eq. Guinea	NT	II
Galago alleni	Allen's Squirrel Bushbaby	Cameroon✪, Eq. Guinea	NT	II
Galago gabonensis	Gabon Squirrel Bushbaby	Cameroon, Gabon✪	NT	II
Galago gallarum	Somali Lesser Bushbaby	Kenya✪	NT	II
Galago matschiei	Spectacled Bushbaby	Uganda✪	NT	II
Galago moholi	Southern Lesser Bushbaby	Botswana✪, Malawi✪, Namibia✪, South Africa✪, Tanzania✪	LC	II
Galago senegalensis	Senegal Lesser Bushbaby	Senegal✪ to Kenya✪, Ghana✪	LC	II
Galagoides cocos	Kenya Coastal Bushbaby	Kenya✪, Tanzania✪	NE	II
Galagoides demidovii	Demidoff's Bushbaby	Cameroon✪, DRC✪, Eq. Guinea	LC	II
Galagoides granti	Mozambique Lesser Bushbaby	Tanzania	DD	II
Galagoides nyasae	Malawi Lesser Bushbaby	Malawi✪, Nigeria, Uganda	NE	II
Galagoides orinus	Mountain Dwarf Bushbaby	Tanzania	DD	II
Galagoides rondoensis	Rondo Dwarf Bushbaby	Tanzania	EN	II
Galagoides thomasi	Thomas's Dwarf Bushbaby	Cameroon✪, Eq. Guinea, Gabon, Ivory Coast, Nigeria, Uganda✪	LC	II
Galagoides zanzibaricus	Zanzibar Bushbaby	Tanzania✪	NT	II
Otolemur crassicaudatus	Thick-tailed Greater Bushbaby	Malawi, South Africa✪, Tanzania✪, Zimbabwe	NT	II
Otolemur garnettii	Garnett's (Small-eared) Greater Bushbaby	Kenya, Tanzania	LC	II
Otolemur monteiri	Silver Greater Bushbaby	Angola to Tanzania✪ and Kenya✪	LC	II

RED LIST: CR = Critically Endangered EN = Endangered VU = Vulnerable NT = Near Threatened LC = Least Concern DD = Data Deficient NE = Not Evaluated ✪ = Best place to watch

Introducing Lemurs

Imagine a bend in a wide, fast-flowing tropical river. The current undercuts a forested bank until, one day, a tangle of trees and vines crashes into the river to be swept out to sea. Clinging to the branches or sleeping in a hollow trunk is a family of primates, which somehow survives until its raft washes up on the shore of an island. This may sound like a highly unlikely film plot, but it is actually the most likely hypothesis to explain how primates first colonized Madagascar.

Mysterious sunken continents, ancient land bridges and changes in sea level have all been proposed to explain the peculiar natural history of Madagascar, the second-largest island in the world. Geological evidence shows that Madagascar became isolated approximately 88 million years ago, long before primates evolved. Once the ancestral lemurs arrived in Madagascar—perhaps as long as 60 million years ago—a wealth of ecological opportunities awaited them. The prosimians they left behind in mainland Africa were mostly displaced as more successful monkeys and apes evolved, but the proto-lemurs had little competition in Madagascar. Over millions of years they evolved to exploit the vast array of unoccupied niches available, resulting in a huge variety of lemurlike primates.

Lemur diversity

Lemurs display extraordinary diversity. Some species are active in the daytime, some in the night, and some in both. Their preferred foods range from fruit and leaves, to bamboo shoots, nectar and insects. Most species are arboreal, spending their lives in trees. Social arrangements in lemurs also vary considerably; some have a largely solitary lifestyle, others pair with a mate for life, and some species form large, complex social groups. Nearly 80 species of lemurs are known, but at least 16 of these have been driven to extinction since humans arrived in Madagascar some 2,000 years ago. The surviving species have many strange adaptations. Being isolated for so long has led to the development of both physical characteristics and behavioral traits found in no other primates. Female dominance, for example, is typical of almost all lemurs, but this trait is seen in very few other primates. Strictly seasonal breeding and activity spread throughout the entire 24-hour period (known by biologists as cathemerality) are also features seen only in lemurs. In fact, some traits of lemurs are unique among mammals.

A number of theories have been put forward to explain the development of such unusual traits. One theory argues that Madagascar's scarce resources and unpredictable environment place extra pressure on reproducing females, and that female lemurs having priority access to resources would therefore help to reduce this pressure. Another theory suggests that such behavioral features may have arisen only in the past 1,000 years, since a lot of species (including predators) have been eradicated by humans relatively recently; the absence of these factors might have allowed some species to become more active during the day.

Above The Red-ruffed Lemur (*Varecia variegata rubra*), a subspecies of the Ruffed Lemur, uses its thick tail to steady itself while foraging among branches in northeastern Madagascar. Fruit makes up about 75 percent of its diet.

Since the arrival of humans in Madagascar, they have exploited the island's resources to devastating effect. Slash-and-burn agriculture is destroying lemur habitat at an alarming rate, as is the cutting of trees. Many lemurs are hunted for meat or caught as pets. The superstitions concerning the hunting of lemurs—beliefs that had protected these primates in the past—have become eroded. Furthermore, climate change threatens lemur habitats. This is a particular threat for endemic island-dwellers such as lemurs, because they have nowhere else to go. Fortunately, the island is the focus of many international conservation efforts. Madagascar is viewed as one of the highest conservation priorities on Earth.

Above The Pygmy Mouse Lemur weighs on average just 1 oz. (30 g).

Dwarf and Mouse Lemurs

Scientists have recently uncovered a staggering diversity of dwarf lemurs in the family Cheirogaleidae, which has grown steadily to include no fewer than 22 species. Among them are the world's smallest primates—the tiny mouse lemurs.

It is difficult to imagine a primate with a head about the size of a human thumb, but in Madagascar evolution has sprung some surprises. Primates don't come more compact than the mouse lemurs. They are not only the smallest primates; they are also the only ones known to estivate (become dormant in summer). Some animals become dormant in order to survive cold conditions (hibernation), but for these tiny lemurs, estivation is a way of coping with food scarcity during the dry season. This state of decreased activity lasts for up to seven months of the year.

During the months leading up to estivation, most dwarf and mouse lemurs gorge themselves, increasing their body weight by as much as 40 percent. Fat-tailed Dwarf Lemurs, as their name suggests, store this extra fat in their tail, which swells massively to look like a big sausage. There are some exceptions to this rule: greater mouse lemurs in the genus *Mirza* do not estivate; and conditions in eastern Madagascar are less variable and extreme than in the west, so it follows that dormancy is less common in eastern species. During the rest of the year, the dwarf, mouse and fork-crowned lemurs that make up the Cheirogaleidae family are busy scampering around the forests of Madagascar. They are able to leap between branches, although most seem to prefer quadrupedal locomotion.

Diet and behavior

The family's feeding habits are fauni-frugivorous, meaning their diet contains fruit, flowers and insects in varying amounts. Due to their fondness for fruit and flowers, they play an important role in the forest as seed dispersers and pollinators. All of these lemurs vary their diet depending on what is available in different seasons. Fork-crowned and Hairy-eared Dwarf Lemurs eat lots of tree gum and other discharged substances, known as exudates, scraping them from the bark of trees using the tooth-comb (a clustered formation of lower canine and incisor teeth), just as bushbabies do. Their long, narrow tongues help to obtain the sticky exudates, while specialized bacteria in the cecum (a chamber in the gut) help to digest it. Exudates also make up a large proportion of the diet of greater mouse lemurs, although usually of the kind secreted by insects rather than trees. The smaller mouse lemurs feed more on insects. An exception is the Mittermeier's Mouse Lemur: its small size would suggest that it should feed on insects, but it mostly eats fruit. The larger dwarf lemurs are not as carnivorous as their smaller counterparts, preferring fruit and nectar.

Having woken up after sunset, dwarf and mouse lemurs set out to forage alone, and any social interactions occur later in the night. Since they are nocturnal animals, sound and smell are more important than vision to their social life. Their vocalizations include

PRIMATE PROFILE
DWARF AND MOUSE LEMURS

SIZE Head and body length: 5–11 in. (12.5–27.5 cm); weight: 1–17 oz. (30–470 g)

APPEARANCE Short, dense fur; long body, short arms and legs; coat color varies between species; large eyes

HABITAT Wide range of moist and dry forest habitats

DIET Fruit, gums, leaves, flowers, buds, seeds, insects (and their exudates), small vertebrates

LIFE HISTORY Poorly known

BEHAVIOR Tree-living, nocturnal, solitary and group foraging; some species have complex social structures

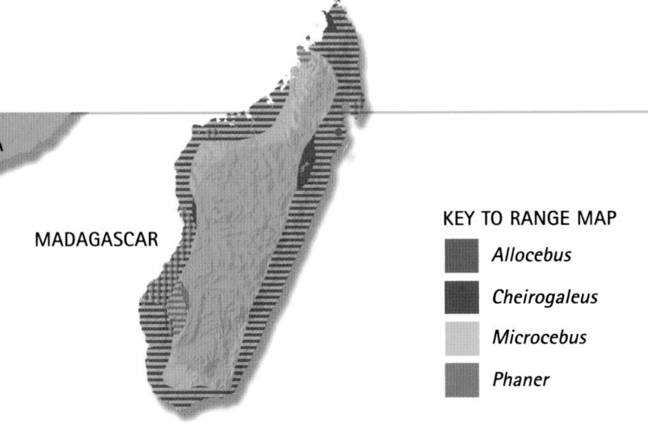

squeaks, whistles and grunts; their olfactory communication involves rubbing urine on the hands and feet to leave a scent trail, and smearing feces on branches while moving. During the day, most species sleep in bundles of a few individuals, inside a nest or tree hole, although the Pygmy Mouse Lemur often prefers to sleep on its own. Male Gray Mouse Lemurs also like to sleep alone, whereas the females snuggle up with as many as 15 others. It appears that most species either breed in separate pairs (monogamous breeding) or use a system in which one male mates with several females (polygynous breeding). Mothers park their young in a safe place when feeding, instead of carrying them around on their back.

Protection

Dwarf and mouse lemurs are the most abundant and widely distributed lemurs and are often spotted in gardens or roadside vegetation. They are also highly adaptable, small and nocturnal—all factors that help to protect them from human persecution. However, as is the case with all lemurs, habitat destruction and hunting today increasingly threaten their survival. The role of these tiny primates in forest ecology is poorly understood, but probably very important, which is why they urgently need to be conserved.

The number of newly discovered mouse and dwarf lemur species highlights just how much there is still to learn about Madagascar and its wildlife. There is a huge amount of work to be done to understand these new species, and with more likely to come to light in the next few years, this is an exciting time for scientists. Madagascar's complex topography presents many geographical barriers to such tiny primates, and this has resulted in the formation of new species between separated populations. However, if forests continue to be destroyed, there is a risk that some species will be destroyed before we even find them.

MAIN DWARF AND MOUSE LEMUR SPECIES

Scientific name	Common name	Where they live	Red List	CITES
Allocebus trichotis	Hairy-eared Dwarf Lemur	Eastern Madagascar	EN	I
Cheirogaleus adapicaudatus	Southern Dwarf Lemur	Southern and western Madagascar	NE	I
Cheirogaleus crossleyi	Crossley's Greater Dwarf Lemur	Eastern Madagascar	NE	I
Cheirogaleus major	Greater Dwarf Lemur	Eastern Madagascar	LC	I
Cheirogaleus medius	Fat-tailed Dwarf Lemur	Western Madagascar✪	LC	I
Cheirogaleus minisculus	Lesser Iron-gray Dwarf Lemur	Central Madagascar	NE	I
Cheirogaleus ravus	Large Iron-gray Dwarf Lemur	Eastern Madagascar	NE	I
Cheirogaleus sibreei	Sibree's Dwarf Lemur	Eastern Madagascar	NE	I
Microcebus berthae	Berthe's Mouse Lemur	Eastern Madagascar	NE	I
Microcebus griseorufus	Gray-brown Mouse Lemur	Southeastern Madagascar	NE	I
Microcebus mittermeieri	Mittermeier's Mouse Lemur	Eastern Madagascar	NE	I
Microcebus murinus	Gray Mouse Lemur	Eastern Madagascar✪	LC	I
Microcebus myoxinus	Pygmy Mouse Lemur	Eastern Madagascar	EN	I
Microcebus ravelobensis	Golden-brown Mouse Lemur	Central eastern Madagascar✪	EN	I
Microcebus rufus	Brown Mouse Lemur	Eastern Madagascar✪	LC	I
Microcebus sambiranensis	Sambirano Mouse Lemur	Northern Madagascar	NE	I
Mirza coquereli	Coquerel's Greater Mouse Lemur	Western Madagascar✪	VU	I
Phaner furcifer	Fork-crowned Lemur	Northeastern Madagascar	LC	I

RED LIST: CR = Critically Endangered EN = Endangered VU = Vulnerable NT = Near Threatened LC = Least Concern DD = Data Deficient NE = Not Evaluated ✪ = Best place to watch

Sportive Lemurs

Sportive lemurs are among the most widely distributed lemurs in Madagascar and have been the focus of much research in the past decade. They are nocturnal and difficult to tell apart, but they have proved to be an ideal subject for testing models and theories of how mammals distribute themselves, and how new species are formed. The results of this research have been astonishing, with 16 new species revealed in 2006–2007 alone.

Until recently, it was thought that there were only one or two species of sportive lemur, split into a total of six subspecies. Early classification relied on external characteristics to distinguish between species, but this was not a reliable method because in closely related nocturnal species, the differences in external features tend to be minor. After all, there isn't much point in an animal wearing a colorful coat when it goes out only after dark. Looks can be deceiving, however, and advances in molecular-level classification techniques have recently enabled scientists to reveal the extensive diversity of animal groups such as the sportive lemurs. To begin with, the subspecies were elevated to full species status, making a total of seven species; then, studies of DNA collected from sportive lemurs from all over Madagascar revealed that even this was an over-simplification, and 11 new species were described during 2006. In 2007, the total stood at 22 species.

AFRICA

MADAGASCAR

KEY TO RANGE MAP

■ *Lepilemur*

Efficient digestion

Sportive lemurs can be found in almost all evergreen and deciduous forests across Madagascar, traveling between the trees by vertical clinging and leaping. Their long, strong back legs enable them to leap from one tree, twist around in midair, then land on the next tree in an upright clinging posture. They eat mainly leaves, which are relatively energy-poor, and have evolved a large cecum (a part of the intestine) to extract the maximum nutrition from this diet. Some researchers have also reported observing sportive lemurs redigesting their feces, a behavior known as cecotrophy. This habit is best known in herbivores such as rabbits and beavers, which produce soft feces on the first passage through the digestive system, then hard pellets the second time around; this behavior allows further digestion of partially processed plant material, enabling maximum extraction of energy from the food. Besides leaves, sportive lemurs may also eat fruit, flowers and seeds, although in much smaller quantities. One study concluded that sportive lemurs cope with their poor-quality

MAIN SPORTIVE LEMUR SPECIES

Scientific name	Common name	Where they live	Red List	CITES
Lepilemur dorsalis	Gray-backed Sportive Lemur	Northern Madagascar	EN	I
Lepilemur edwardsi	Milne-Edwards' Sportive Lemur	Northwestern Madagascar✪	EN	I
Lepilemur leucopus	White-footed Sportive Lemur	Madagascar	EN	I
Lepilemur microdon	Small-toothed Sportive Lemur	Eastern Madagascar✪	EN	I
Lepilemur mustelinus	Weasel Sportive Lemur	Northeastern Madagascar	EN	I
Lepilemur ruficaudatus	Red-tailed Sportive Lemur	Southwestern Madagascar	EN	I
Lepilemur septentrionalis	Northern Sportive Lemur	Northern Madagascar✪	EN	I

RED LIST: CR = Critically Endangered EN = Endangered VU = Vulnerable NT = Near Threatened LC = Least Concern DD = Data Deficient NE = Not Evaluated ✪ = Best place to watch

diet simply by resting more and traveling less during the cooler season, thereby conserving energy. It was long believed that most if not all sportive lemurs forage alone at night. However, the Milne-Edwards' Sportive Lemur has since been seen feeding in groups of up to three individuals, with group members taking regular breaks to groom each other. This discovery mirrors recent research, which has revealed greater complexity in prosimian social systems.

Social habits

The social and mating systems of sportive lemurs are poorly understood but, as research continues, more will doubtless be revealed. Some species are still thought to lead entirely solitary lives, but others seem to live in dispersed pairs, with a male and female sharing a tree hole during the day and separating at night to forage. Pairs of Milne-Edwards' Sportive Lemurs will actively defend their shared territory by branch-shaking and vocal displays. Sportive lemurs have an elaborate vocal repertoire, including crowlike "loud calls" that are used to defend territory. They are threatened, like all lemurs, by habitat destruction and hunting. However, evolving research is giving a greater insight into these prosimians and the dark forests in which they live, which should eventually lead to better protection for these fascinating primates.

PRIMATE PROFILE
SPORTIVE LEMURS

SIZE Head and body length: 12–14 in. (30–35 cm); weight: 1 1/8–2 lb. (500–900 g)

APPEARANCE Short, dense brown or gray fur; thin tail; large eyes and muzzle

HABITAT Wide range of moist and dry forest habitats

DIET Fruit, leaves, flowers, seeds, bark

LIFE HISTORY Gestation: 4–5 months; sexual maturity: 18–21 months; life span: not known

BEHAVIOR Active at night; tree-living, highly territorial with well-defined and well-defended (by males) home ranges of about 2 1/2 acres (1 ha); usually solitary

Right Most sportive lemurs are solitary, but some, such as these Milne-Edwards' Sportive Lemurs, share a sleeping hole with a mate. When they emerge at night, they may forage alone or together in the trees.

True Lemurs, Including Bamboo Lemurs

The members of the Lemuridae family form a large and diverse group of species known as true lemurs, and their classification is a subject of much scientific debate. Over the years, several lemur forms have been upgraded to full species, downgraded, presumed extinct and then discovered again, and today the family may contain up to 20 species, depending on which view you favor. This family comprises the familiar Ring-tailed Lemur, the charismatic Ruffed Lemurs, the many subspecies of Brown Lemurs and the highly specialized bamboo lemurs.

Above True to its name, the male Black Lemur (left) has a black coat. The female, however (right), has a brown back and white ear tufts. This species is active by day and night and feeds upside-down on the tips of branches, foraging in groups

Main true lemur species

The Ring-tailed Lemur is one of the best-studied and most frequently filmed of all primates. But despite being well known, this lemur has an uncertain taxonomy and its exact position on the lemur branch of the primate family tree has been much debated. It shares some behavioral traits with the bamboo lemurs and also with the many *Eulemur* species, which were originally classified in the same genus. Ring-tails now stand alone in the genus *Lemur*. The *Eulemur* genus, in contrast, is large and highly diverse, containing as it does the many different subspecies of Brown Lemur (recent studies suggest some may even be species in their own right). Brown Lemurs are very flexible in both diet and habitat and therefore are widely distributed across Madagascar. In contrast, the bamboo, or gentle, lemurs in the *Hapalemur* genus have specialized in feeding on bamboo. Last, but certainly not least, the Lemuridae includes the beautiful Ruffed Lemur, which is renowned for its loud, raucous calls.

Most members of the Lemuridae organize themselves into multimale–multifemale groups that contain anything from 3 to 24 members. The Ring-tailed Lemur forms particularly large and complex groups with strict social hierarchies, but unlike the other species, a typical group of Ring-tails consists of a dominant adult female and several males that will later disperse between various other groups when they reach three to four years of age. The Red-bellied Lemur is one of the few monogamous lemurs, in which bonded pairs actively defend a territory, although Mongoose and Ruffed Lemurs have occasionally been observed in male–female pairs as well as in the more usual mixed-sex groups.

One of this family's many fascinating features is different coloration between the sexes, known as sexual dichromatism. It can be seen to varying degrees in each species, and is at its most extreme in the Black Lemur (see left). The color differences in other species are more subtle—often being restricted to facial markings—while the Common Brown Lemur subspecies (*Eulemur fulvus fulvus*) shows no sexual color difference at all. Interestingly, sexual dichromatism appears to be linked to the degree of female

dominance. Female dominance is less obvious in Brown Lemurs and their many subspecies, which happen to be the forms with the smallest color difference between female and male.

Most of the Lemuridae feed mainly on fruit. They also eat leaves, flowers, nectar and insects in varying, but smaller, amounts. The Mongoose Lemur eats mainly nectar during the dry season, and may play an important role as a pollinator, as may many other lemurs due to their taste for nectar and flowers. By eating much fruit and traveling around the forest, lemurs also spread the seeds of many important tree species. This is vital for forest regeneration and shows how protecting lemurs from extinction benefits humans, because we need the forests to maintain the health of the planet.

Varied habitats

These lemurs occupy many different habitats across Madagascar, from dry scrub in the south to mountainous rainforest in the east and coastal forests all around the island. The Alaotran Bamboo Lemur subspecies (*Hapalemur griseus alaotrensis*) has filled a highly unusual ecological niche for a primate—it lives in the marshy reedbeds surrounding Lake Alaotra in the northern central plateau. Mongoose Lemurs are the only lemur found away from mainland Madagascar, with a population on the nearby Comoros Islands, but this was probably established by human introductions. All species in the family live in trees, although the Ring-tailed Lemur (see left) may be best described as semiterrestrial because it spends up to a third of its time on the ground.

Left Ring-tailed Lemur females are receptive to mating for only 24 hours in April each year, then give birth to a single infant in late August. The female carries the infant on her back for the first two weeks of its life.

PRIMATE PROFILE
TRUE LEMURS

SIZE Head and body length: 11–18 in. (28–46 cm); weight: 2½–10 lb. (700 g–4.5 kg)

APPEARANCE Thick, woolly fur (color varies between species and sometimes between sexes); long, bushy tail

HABITAT Scrub and desert; dry and humid forest; bamboo; plantations

DIET Fruit, leaves, seeds, nectar, stems, bamboo, small vertebrates, insects.

LIFE HISTORY Gestation: 4–5 months; sexual maturity: 20–30 months; life span: up to 27 years

BEHAVIOR Active at night, during the day, or during both day and night, according to species; social structure varies; highly vocal; scent used to mark territory

If you see a Ring-tailed Lemur, it is likely to be parading on the ground with its striking piebald tail held high. Most members of the Lemuridae family travel through their habitats by walking, running and leaping using all four limbs.

Scent and communication

Scent plays a big part in lemur social life. All of the Lemuridae secrete scent from glands underneath the tail, some species applying it to trees by doing handstands. *Eulemur* species also have scent glands on their heads and palms, and Ring-tailed Lemurs have spurs on their wrists that they use to scratch bark on trees before marking them. Ring-tailed and bamboo lemurs use scent in another way too—a curious behavior known as stink fighting. When engaged in these smelly battles, male Ring-tails draw their tails up between their legs and through their wrist glands, then wave the stink-adorned stripy tail in the direction of their opponent. This behavior is unique to Ring-tailed and bamboo lemurs, suggesting that they may be closely related. Vocal communication is important in several species of Lemuridae. Ruffed Lemurs are, along with Ring-tails, some of the most vocal. The former have an elaborate system of alarm calls and their loud, raucous barking is highly contagious, spreading through the forest if other groups are living nearby.

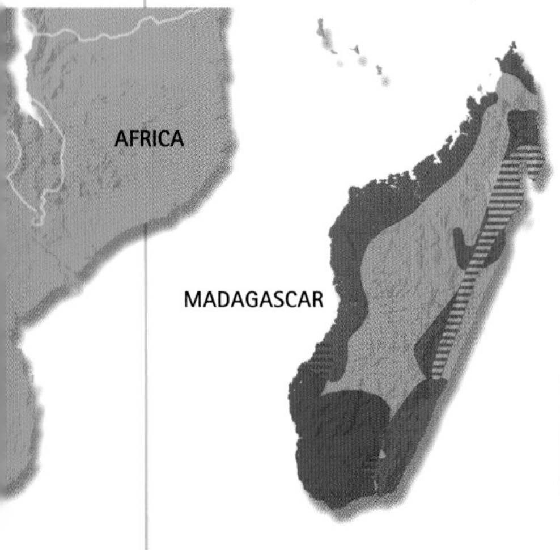

AFRICA

MADAGASCAR

KEY TO RANGE MAP

Eulemur

Lemur

Varecia

Family life

Lemurs in this family usually give birth to single infants, or occasionally twins, if conditions are good. An exception to this rule is the Ruffed Lemur, which is the only diurnal primate known to give birth to a litter and then keep the young in a nest; litters of between two and four young are the norm, although not all will survive. Young Ruffed Lemurs are not carried like those of other Lemuridae, but are stashed somewhere safe high up in the canopy while the mother forages. Other members of the Ruffed Lemur group may also help to look after the litter, guarding, stashing, transporting and even nursing the young lemurs.

Many lemurs in the Lemuridae spread their activity throughout the entire 24-hour period. This phenomenon is a characteristic essentially restricted to the primates of Madagascar, and much debate and research has surrounded its functions and origins in recent years. Relative amounts of day and night activity seem to be influenced by different factors, including seasonality, phases of the moon and changing light levels.

All members of this diverse family are threatened. Perhaps the most endangered is the stunning Blue-eyed Black Lemur subspecies (*Eulemur macaco flavifrons*), the only primate besides humans to have truly blue eyes. The enigmatic Red-ruffed Lemur subspecies (*Varecia variegata rubra*) is also at grave risk due to its extremely restricted range. Moreover, the large fruit trees in which Ruffed Lemurs feed are also the ones that loggers selectively fell because they are the most valuable.

Bamboo lemurs

Bamboo, or gentle, lemurs are the only primates anywhere that feed almost entirely on bamboo, making them highly unusual even by lemur standards. In the giant bamboo forests of east-central Madagascar, you can soon spot where a bamboo lemur has been

TRUE, RING-TAILED AND RUFFED LEMUR SPECIES

Scientific name	Common name	Where they live	Red List	CITES
Eulemur coronatus	Crowned Lemur	Northern Madagascar✪	VU	I
Eulemur fulvus	Brown Lemur	Madagascar✪	LC	I
Eulemur macaco	Black Lemur	Northwestern Madagascar✪	VU	I
Eulemur mongoz	Mongoose Lemur	Madagascar✪	VU	I
Eulemur rubriventer	Red-bellied Lemur	Madagascar✪	VU	I
Lemur catta	Ring-tailed Lemur	Southern Madagascar✪	VU	I
Varecia variegata	Ruffed Lemur	Eastern Madagascar✪	EN	I

RED LIST: CR = Critically Endangered EN = Endangered VU = Vulnerable NT = Near Threatened LC = Least Concern DD = Data Deficient NE = Not Evaluated ✪ = Best place to watch

Above Lesser Bamboo Lemurs are diurnal (active during the day) and crepuscular (active at dawn and dusk). They spend nearly half their time feeding.

feeding—half-eaten bamboo shoots litter the forest floor. Follow the trail and listen carefully—if you hear the sound of eager tearing of vegetation and rapid munching, you might just be in luck. Sitting among the thick vegetation may be a bamboo lemur, fervently ripping up bamboo shoots from all around itself and tossing aside the parts it prefers not to eat, rather like a miniature Giant Panda.

Poisonous diet

Superb dexterity and hand-eye coordination enable this lemur to spot a tasty bamboo shoot and then grasp, strip and devour it in the time it takes the average lemur simply to sniff out its next morsel. Different species of bamboo lemur prefer to eat different parts of the bamboo plant, which contains high levels of poisonous cyanide, so they have clearly developed a mechanism for withstanding extremely large amounts of this deadly chemical. Certain species, such as the Greater Bamboo Lemur, favor particular species of bamboo. Being even more specialized in diet limits the areas in which this species can live, and this unfortunately makes it more vulnerable to habitat destruction. When a bamboo lemur has eaten its fill, it may move on to a resting spot, traveling through the vegetation by leaping from tree to tree in an upright position. Bamboo lemurs differ from other lemurs that use this "vertical clinging and leaping" style of locomotion, because they also often travel on all fours on the ground or along horizontal branches. Amazingly, the Alaotran Bamboo Lemur (*Hapalemur griseus alaotrensis*), which lives only in the reed beds on Lake Alaotra, Madagascar, also regularly swims in its reedbed home, while out foraging for bamboo.

ICA

MADAGASCAR

KEY TO RANGE MAP

- *Hapalemur aureus*
- *Hapalemur griseus*
- *Hapalemur simus*

Parking the young

Bamboo lemurs live in groups numbering between two and nine. These social groups may consist of one or two breeding females and their offspring, plus a breeding male. Usually only single infants are born, and they are "parked" on a nearby branch while the mother forages. She will return to her infant at intervals, to groom and nurse it. This system of "infant stashing" differs from the parental care seen in most other lemurids, which usually involves carrying the young on the mother's back. Young bamboo lemurs become mobile at a surprisingly young age. Most lemurs in this genus, like other lemurs, breed seasonally. The Alaotran Bamboo Lemur is the exception—it will mate in any season.

If bamboo lemurs encounter another group of the same species within their home range, they defend their territory by chasing the intruders, scent-marking and vocalizing. Males will also engage in "stink fights" (*see* facing page). When a bamboo lemur group becomes separated, a lone individual might produce a *"coooee"* call, while alarm calls include *"co-ouiiiii"* and *"graaaa"* sounds. Other vocalizations include grunts and teeth-grinding.

Much like the Giant Panda in China, the bamboo lemurs' strict dietary requirements make them highly vulnerable to change. Species that have further specialization for specific bamboo species and plant parts, or restricted ranges and habitat requirements, are at even more risk. Ensuring the protection of these and other lemur habitats is vital now more than ever, before populations fall below levels from which they can recover.

MAIN BAMBOO LEMUR SPECIES

Scientific name	Common name	Where they live	Red List	CITES
Hapalemur aureus	Golden Bamboo Lemur	Eastern and western Madagascar✪	CR	I
Hapalemur griseus	Lesser Bamboo Lemur	Eastern Madagascar✪	CR	I
Hapalemur simus	Greater Bamboo Lemur	Eastern Madagascar✪	CR	I

RED LIST: CR = Critically Endangered EN = Endangered VU = Vulnerable NT = Near Threatened LC = Least Concern DD = Data Deficient NE = Not Evaluated ✪ = Best place to watch

The Indrid Family

Above A young Verreaux's Sifaka needs to cling on tightly as its mother bounds sideways in classic sifaka style across a clearing to the safety of a tree.

The Indri's spine-tingling song and the sifakas' acrobatic dancing make this beautiful family the most spectacular group of lemurs. Indris and sifakas are the largest living prosimians and the heaviest animals to move by vertical clinging and leaping. The indrid family also includes the secretive nocturnal woolly lemurs, about which much remains to be discovered.

Throughout Madagascar, numerous legends are told of the origins of the mysterious Indri, or *babakoto*, meaning "ancestor of men," all involving the theme of common ancestry. In contrast to the Aye-aye (*see* pages 70–71), which is regarded as an omen of death by the Malagasy, this lemur is held to be sacred, and to kill one is said to bring misfortune.

Dueting lemurs

The Indri's songs are sung in duet by monogamous pairs of adults that form the social group; if offspring are present, they too join in the first part of the chorus. The call spreads from one group to the next across the mountain, each call carrying up to about 2 miles (3 km), until every Indri group in the forest has declared its location using its own unique variation of the song.

Such "long calling" uses less energy than other types of territorial marking, such as scent-marking, and energy is at a premium for Indris due to the low nutritional value of the leaves that make up the majority of their diet. The woolly lemurs in the genus *Avahi* spend up to 60 percent of the time resting to avoid wasting precious energy. On the other hand, the sifakas (pronounced *she-faks*) eat more high-energy fruit, so they can afford to invest more energy in producing and laying down sebaceous (oily) and apocrine (sweaty) scent marks to communicate with each other. Vocal communication is still important in sifakas and woolly lemurs, however. Woolly lemurs produce an *"ava-hy"* noise when alarmed, while the sifakas make a curious *"shi-fak"* sound.

Indrids have a surprising method of locomotion for such big animals, quite unlike that of other lemurs. Using their long, powerful back legs, they spring from one vertical support to the next, making leaps of up to 30 ft. (9 m) between trees, while maintaining an upright position. When traveling across the ground, they switch to a spectacular upright skipping motion—a behavior best seen in the sifakas (*see* page 49). A sifaka will skip sideways across open areas with its tail outstretched and arms flailing in the air to keep balance, a sight that is breathtaking to watch.

Social organization varies between the species. Indris, woolly lemurs and Diademed Sifakas form monogamous pairs. But Verreaux's Sifakas form groups of between two and nine individuals, usually with more males than females in the group.

Classification of indrids

The complex taxonomy of the indrids is still being unraveled. There are two subspecies of Indri, differing slightly in their coloration and distribution. Four species of night-active woolly lemurs are currently known, each with a different habitat preference. One of them, the Bemeraha Woolly Lemur, was first sighted by scientists as recently as 1990 and formally described in 2005, when it was named *Avahi cleesei* after the British actor John Cleese. The sifaka genus, *Propithecus*, contains at least three species, which are further divided into eight subspecies, some of which may be species. There is some debate concerning the form known as Milne-Edwards' Sifaka: one genetic study found it to be a species in its own right, but others concluded that it is a subspecies of the colorful Diademed Sifaka, which is already split into three other subspecies. Sifakas are regarded by many to be the most beautiful of Madagascar's lemurs, with their long, silky

PRIMATE PROFILE
INDRIDS

SIZE Head and body length: 10–28 in. (25–70 cm); weight: 2½–18 lb. (700 g–8 kg)

APPEARANCE Short, dense fur (color varies between species); long, powerful legs; webbed toes; bare muzzle

HABITAT Dry and humid forest; scrub

DIET Fruit, leaves, seeds, buds, flowers, bark

LIFE HISTORY Gestation: 4–5 months; sexual maturity: 4–7 years; life span: not known

BEHAVIOR Active at night or during the day, according to species; social structure varies from monogamous pairs and their young to groups of up to nine adults; distinctive calls define territorial boundaries

hair and hairless black faces. One of the most recently discovered diurnal lemurs is the Golden-crowned Sifaka, which was also thought to be a form of Diademed Sifaka until genetic studies revealed otherwise. With an estimated 6,000–10,000 individuals left in the wild, this species is one of the top 10 most endangered primates in the world. A key reason for this is its extremely limited distribution: its habitat is very fragmented and survives in a severely restricted range that is only 22 miles (35 km) across at its widest point, which has led some groups to become isolated from the main population.

Although cultural taboos hold the Indri to be sacred, growing numbers of newcomers to Madagascar do not hold such beliefs, and the Indri is now more threatened than ever before.

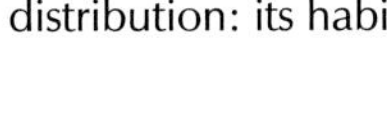

MADAGASCAR

KEY TO RANGE MAP

- *Avahi*
- *Indri*
- *Propithecus*

Above The Indri is large with a very short tail. Among the Betsimisaraka people of eastern Madagascar this species is known as *babakoto*, which loosely translates as "grandfather" or "ancestor of men."

INDRID SPECIES

Scientific name	Common name	Where they live	Red List	CITES
Avahi cleesei	John Cleese's Lemur	Western Madagascar	NE	I
Avahi laniger	Eastern Woolly Lemur	Madagascar	LC	I
Avahi occidentalis	Western Woolly Lemur	Northwestern Madagascar	VU	I
Avahi unicolor	Unicolor Avahi	Northwestern Madagascar	NE	I
Indri indri	Indri	Northeastern Madagascar ✪	NE	I
Propithecus diadema	Diademed Sifaka	Eastern Madagascar ✪	EN	I
Propithecus tattersalli	Golden-crowned Sifaka	Northern Madagascar	CR	I
Propithecus verreauxi	Verreaux's Sifaka	Western Madagascar ✪	VU	I

RED LIST: CR = Critically Endangered EN = Endangered VU = Vulnerable NT = Near Threatened LC = Least Concern DD = Data Deficient NE = Not Evaluated ✪ = Best place to watch

Aye-aye

So strange that it baffled taxonomists for years, the Aye-aye was first classified as a rodent, then finally given its own primate family: Daubentoniidae. Its bizarre appearance and feeding habits make it probably the world's most unusual primate. Perhaps unsurprisingly, the people of Madagascar have long considered this weird-looking creature to be an omen of death.

The Aye-aye is the largest of the nocturnal primates. Its peculiar features have evolved to fit a specialized ecological niche usually occupied by woodpeckers and squirrels, both of which are absent from Madagascar. Huge "gremlin" eyes sit within the Aye-aye's ferretlike face, while its long and shaggy coat of white-tipped dark hairs give it a rather unkempt appearance. The paler face, with dark rings circling the eyes, is topped with huge black batlike ears capable of independent rotation. The animal's total length is almost doubled by the big, bushy tail, which is similar to a squirrel's, the hairs on which are the longest found in any prosimian. The Aye-aye's dentition is unique among primates—it lacks canine teeth, has small cheek teeth and possesses a single

the Aye-aye is equipped with a very long, very thin third finge which it is able to rotate through 360°, independently of the other, shorter fingers.

When hunting at night, the Aye-aye moves along a branch rapidly tapping the wood with its elongated fingers while listening intently with its huge ears cupped. If it detects the hollow sounc of a cavity deep within the tree and hears an insect larva moving inside, it starts to gnaw, sending bits of wood flying in all directions until it is able to insert its long, bony third finger to extract th

Above Resembling a cross between a bat and a giant squirrel, the nocturnal Aye-aye, with its bizarre, elongated middle finger, strikes fear into the minds

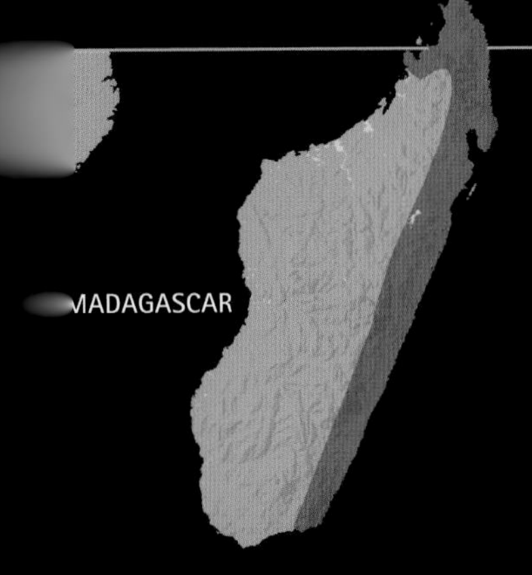

KEY TO RANGE MAP

Aye-aye

PRIMATE PROFILE
AYE-AYE
Daubentonia madagascarensis

SIZE Head and body length: 16 in. (40 cm); tail length: 16 in. (40 cm); weight: male 6 lb. (2.8 kg), female 5¾ lb. (2.6 kg)

APPEARANCE Coarse, patchy, black-brown fur, edged in white; pale face; pointed nose; dark eye rings; large, hairless ears; elongated middle fingers; long, bushy tail

HABITAT Rainforest; dry forest; plantations; spiny desert

DIET Insect larvae, seeds, fruit, nectar

LIFE HISTORY Gestation: 6 months; sexual maturity: female's age at first birth 3–4 years; life span: not known

BEHAVIOR Nocturnal; tree-living; solitary; urine and scent marks; loud vocalizations

RED LIST EN **CITES** I

unsuspecting grub. A third eyelid (yet another feature that is unique among primates) protects this primate's eyes from the flying debris. Young Aye-ayes spend much time learning the percussive foraging technique, the first few clumsy attempts of which are made at around 10 weeks old. When an Aye-aye is eating juicy grubs or coconuts, the long finger is used to rapidly scoop out liquid contents.

During the day, Aye-ayes sleep in nests made from twigs woven together and lined with shredded leaves. As they are solitary, the nests are normally occupied by a single animal, although different individuals may use the same nest on different days. Aye-ayes emerge at around sunset to groom, but they sometimes wake up a little earlier. Aye-ayes often hang upside down to groom, with the third finger being used to clean the eyes, ears and nose. The majority of the night is then spent foraging, pausing occasionally to rest and groom some more. Aye-ayes clamber around the highest levels of the forest using four limbs, descending to lower levels to leap between trees. When climbing down a vertical trunk, they will sometimes go head first, gripping with their clawlike nails.

Males cover huge ranges of up to 500 acres (200 ha), which overlap the territories of other males and females. This may be a reproductive strategy, because Aye-ayes have no specific breeding season and therefore each female may be fertile at different times. Female ranges are much smaller, approximately 75–125 acres (30–50 ha), and never overlap with those of other females.

Above An Aye-aye uses its distinctive elongated middle finger to probe for a favored meal of grubs under the bark of a tree trunk.

Reproduction and social behavior

When a female enters a fertile period, she will call to males, of which up to six at a time surround her and fight for the chance to copulate. In the end, she may mate with several males. Although Aye-ayes traditionally have been described as solitary, they actually live in a social matrix, communicating mostly through vocal and olfactory means. Up to four Aye-ayes have been observed feeding near each other. Interactions between females are rare and always aggressive, but the males get along much better. Their vocal repertoire includes *"eeps"* and *"creees"* and a *"ron-tsit"* call to make contact and express distress and alarm. They give a contented *"hoo-hoo"* when feeding on a favorite food or grooming after copulation.

Until humans arrived on Madagascar, another, larger species, the Giant Aye-aye, *Daubentonia robusta*, lived in the dry southwest of the island. Its teeth have been found with holes drilled in them, so it may have been hunted for jewelry by the early settlers. The single surviving Aye-aye species now also faces extinction, due to a combination of habitat destruction and human persecution. Its reputation as a portent of death is not helped by its taste for coconuts and sugar cane, which brings it into conflict with farmers. In 1935, the Aye-aye was declared extinct, only to be rediscovered in 1957. The species is now known to have a wide but sparse distribution in eastern Madagascar. In an attempt to preserve the species, a few Aye-ayes were taken to Nosy Mangabe, an island just off the east coast of Madagascar, where the population has since thrived.

Tarsiers

With their huge, endearing eyes and mobile heads that can swivel 180°, tarsiers look like the creation of a Hollywood special-effects team. They are fascinating, if elusive, primates with a long evolutionary lineage. They possess features of not only prosimians but also monkeys and apes, and they are the subject of much academic debate. However, scientists attempting to study tarsiers in rainforests must first contend with a more prosaic problem—how to locate their research animals.

Above Tarsiers have an almost owl-like ability to swivel their head right around, as demonstrated by this Western Tarsier in the Malaysian province of Sabah on the island of Borneo. They need this ability because they cannot move their eyes.

Tarsiers are agile and fast-moving "vertical clingers and leapers" that can jump extraordinary distances to cross gaps between forest trees. Their leaping ability stems from the powerful legs, which are one-and-a-half times the length of the head and body combined. This is not apparent when a tarsier is at rest because the bones of the thigh, lower leg and foot—all roughly about the same length—are folded in a concertina shape. But the arrangement is remarkable because it gives a triple extension when the animal jumps. On landing, it is the feet of the tarsier that touch the tree trunk first, rapidly followed by both hands to grip the branch or trunk firmly. Tarsiers are named for the tarsus, the bone that forms the ankle in the human foot, and which is greatly elongated in all *Tarsius* species.

PRIMATE PROFILE
TARSIERS

SIZE Head and body length: 34–63 in. (85–160 cm); weight: 3½–5 oz. (100–140 g)

APPEARANCE Soft, dense fur (color varies between species); long tail, partly hairless; very large eyes; short snout; short forelimbs, long hind limbs

HABITAT Dry and humid montane and lowland forest; coastal forest and mangroves; scrub

DIET Small vertebrates, insects

LIFE HISTORY Gestation: 6 months; sexual maturity: 1 year; life span: up to 12 years

BEHAVIOR Tree-living; active at night; some species hide in ground hollows; scent-marking; social structures vary

Large-eyed hunters

Tarsiers are entirely carnivorous primates, catching and eating large insects such as crickets, cockroaches and moths, as well as small vertebrates such as birds, bats and snakes. Prey is seized in both hands and killed with a bite before being crunched up bit by bit in the tarsier's sharp teeth; some observers have noted that a tarsier closes its eyes at the point of catching its prey, presumably to avoid injury. The eyes are larger, relative to the head, than in any other species, with each eyeball weighing more than the whole brain. Unlike most nocturnal animals, tarsiers lack a *tapetum lucidum* (which means "bright carpet"). This is the reflective layer in the back of the eye, behind the retina, that reflects any light not absorbed by the retina, producing the familiar "cat's eyes" reflection in a flashlight or car headlight.

The absence of a reflective layer in the eyes makes tarsiers difficult to find in tangled undergrowth at night, but modern technology, in the form of tiny radio collars used for tracking, has helped to overcome this difficulty and a fuller picture of tarsier society is finally beginning to emerge. Tarsier territories are 7–27 acres (2.8–11 ha) in size, with males usually ranging further than

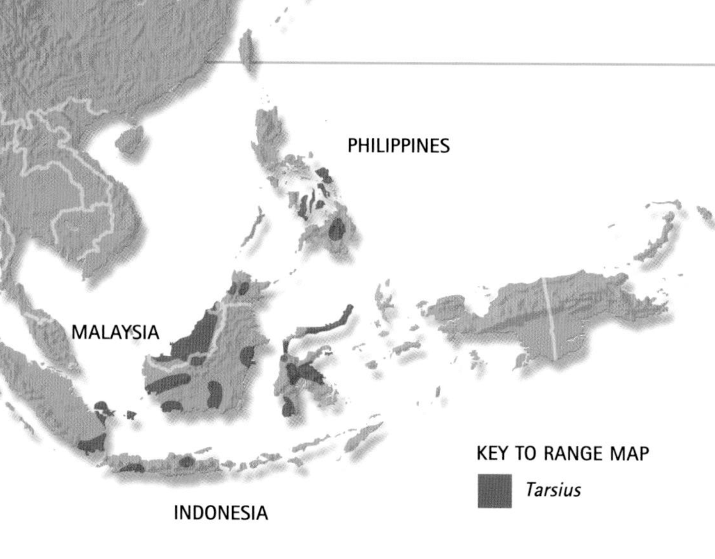

Until the 1980s, only three species of tarsier were known—the Western, Spectral and Philippine tarsiers. However, once field studies began in Sulawesi, it was realized that animals described as the Spectral Tarsier were in fact a number of different species (the exact number is still to be determined). It is likely that when further studies are undertaken, particularly of the tarsiers' DNA and vocalizations, new species will be described.

females. The animals spend more than half the night foraging, a quarter traveling and the rest socializing. Sleeping habits vary: some species return to the same sleeping site—usually a fig tree or a tangle of vines—night after night; others select different trees each night. Most tarsiers sleep singly, although sleeping trees are often found in the areas where neighboring male and female ranges overlap. Spectral Tarsiers, however, live in groups and up to eight may spend the day together at the same sleeping site.

The single infant does not cling to its mother's body, but is left while she forages nearby, typically within about 13 ft. (4 m) of her offspring. It was thought that young tarsiers were left unattended all night, but a study of Spectral Tarsiers showed that the usual length of time the infant was "parked" was only about half an hour. The mother carries her infant in her mouth to a new site 11 times a night on average—this kind of maternal behavior has been termed "cache and carry."

Right Having fixed the location of an insect (in this case a bush cricket) using sight and sound, a Philippine Tarsier, like all tarsiers, can leap extraordinary distances to catch its prey.

TARSIER SPECIES

Scientific name	Common name	Where they live	Red List	CITES
Tarsius bancanus	Western (Horsfield's) Tarsier	Borneo✪, Indonesia	LC	II
Tarsius dentatus	Dian's Tarsier	Sulawesi (Indonesia)	NT	II
Tarsius lariang	Lariang Tarsier	Sulawesi (Indonesia)	NE	II
Tarsius pelengensis	Peleng Tarsier	Peleng Island (Indonesia)	DD	II
Tarsius pumilus	Pygmy Tarsier	Sulawesi (Indonesia)	DD	II
Tarsius sangirensis	Sangihe Tarsier	Greater Sangihe Island (Indonesia)	DD	II
Tarsius syrichta	Philippine Tarsier	Philippines✪	DD	II
Tarsius tarsier	Spectral Tarsier	Indonesia✪	NT	II

RED LIST: CR = Critically Endangered EN = Endangered VU = Vulnerable NT = Near Threatened LC = Least Concern DD = Data Deficient NE = Not Evaluated ✪ = Best place to watch

New World Monkeys

Main image Bald-faced Sakis and their close relatives do not have fully prehensile (grasping) tails, but they can curl their tail loosely around branches for balance high in the forest canopy.

Introduction: New World Monkeys

The tropical forests of South and Central America hold a primatological puzzle. There are no prosimians or apes here, but there is a rich diversity of monkeys quite different from those in Africa and Asia. The origin of these New World primates and the nature of their arrival in the Americas has been debated for years.

Above The Red-bellied Tamarin is found only in a small area on the border between Peru and Brazil. This species spends most of its time high in the forest canopy, more than 33 ft. (10 m) from the ground.

The continents of Africa, North America and even Antarctica have all been suggested as the place where these primates originated. The South American continental plate had long since separated from Africa when they suddenly appeared in the fossil record 27 million years ago. Both land and sea crossings have been put forward as methods of migration to South America, but unlikely as it seems, most of the evidence points to a transatlantic journey from Africa. The suggestion is that early African primates were swept away on a huge raft of floating vegetation that may have broken from the mainland during a terrific storm. Prevailing ocean currents and strong winds took the marooned primates to a new South American home.

These primates would become the platyrrhines, or New World monkeys. Over millions of years, they branched off in all directions as they evolved to exploit different niches in their new world, forming many new species. However, fossils that document this time of change are rare, and many questions remain unanswered. Morever, primates in many different branches on this part of the family tree have evolved in similar ways to adapt to the same environment. This process, known as convergent evolution, makes it difficult to classify species based on their external appearance.

Flat noses

Platyrrhini means "flat nosed," and a flat nose is the most obvious distinguishing feature of these primates. The monkeys of South America have a wide septum that separates sideways-facing nostrils, in contrast to the downward-facing nostrils and narrow septum of Old World monkeys, apes and humans. Their teeth also differ: most New World monkeys have 36, which is 4 more than their Old World cousins, humans included. Marmosets and tamarins are the exception—they have 32 teeth.

Most New World monkeys are relatively small and arboreal, making them more difficult to study than some of their larger, more terrestrial Old World relatives. Many of them have a long, prehensile (able to grasp) tail, forming a well-muscled extra limb for clambering and swinging through the trees, picking up objects or even wrapping around a fellow monkey in a welcoming embrace.

Left Most primates like to occupy a good vantage point, as this Brown Woolly Monkey demonstrates. The best position is usually occupied by the alpha male, so he can watch who is doing what with whom and keep a lookout for danger.

Staggering diversity

The Amazon Basin supports the largest concentration of primate diversity in the world—to date 81 species have been described, with more still being found. This amazing diversity is explained by the area's geological history. Before the break up of continents, the Amazon River was part of the proto-Congo River system, flowing westward through Africa. As the ancient continent of Gondwanaland split apart 130 million years ago, South America brought its river with it when it broke away.

Today, the Amazon Basin contains a wide variety of habitat types, including deciduous, flooded and seasonal forest, rainforest and savanna. The Amazon tropical forest is the largest on Earth, and it ranks alongside Madagascar and Borneo as one of the planet's top biodiversity hotspots. At about 2.1 million square miles (5.5 million sq km) in area, the forest makes up more than half of the remaining rainforest on Earth. The region's species total is continually rising as new forms are discovered; in 2007, for example, 24 new animal species were discovered in Suriname alone. These species are ecologically dependent upon each other, as is the case in all ecosystems, and primates play a particularly important role in the forest ecosystem: by spreading the seeds of the fruit they eat, they act to renovate and diversify the forest.

Threat of change

The tropical forests of the Amazon are so vast that they exert considerable influence over the climate of the entire planet. They absorb huge amounts of carbon, storing about 86 billion tons—equal to 11 years' worth of recent global carbon emissions. Although the current rate of forest loss, almost 4,000 square miles (10,000 sq km) a year, is the lowest since records of deforestation began in 1988, this is still a staggering amount. Owing to human activities, the Amazon forest is one of the fastest changing ecosystems on Earth. Deforestation and climate change are altering water cycles, leading to severe droughts, which in turn trigger forest fires. Some of the worst Amazon droughts and fires on record were in 2005.

Other threats loom over the future of the forests and their species. A huge Amazonian dam has recently been approved, which could spell disaster. A new mining project is underway in Suriname. Illegal logging continues to be a problem, although more readily available satellite imaging is proving an effective way of reducing this. Ultimately, providing local people with incentives to protect the forests on which they depend may be the most effective method of ensuring a future for South America's marvelously diverse primate species.

Right The Common Squirrel Monkey lives in troops of 20 to 40 individuals, which run and jump through the forest trees chattering and chirping, often in association with capuchins and uakaris.

Facing page The Bearded Emperor Tamarin subspecies (*Saguinus imperator subgrisescens*) searches for insects in the lower to middle levels of the forest.

Marmosets and Tamarins

With their thick, lustrous coats, draping manes, spectacular crests, ear tufts and mustaches, the marmosets and tamarins are a very colorful group. There is much more of interest in their lives than hairstyles, however. These monkeys also have many characteristics that are unique among primates, making them fascinating subjects for primatologists.

The subfamily name of these striking primates is Callitrichidae, which literally means "beautiful hair," and they certainly live up to their name. There are about 60 species and subspecies altogether, all with distinctive appearance. They are grouped into six genera.

Above The Pygmy Marmoset is the smallest living monkey at about 6 in. (15 cm) long, excluding the tail. Sap-feeding holes made in tree bark, as seen here, are a sign of its presence.

Arboreal acrobats

A highly acrobatic group, marmosets and tamarins run, bound and leap between the trees. They use their long tails for balance, but cannot grip with the tip as most other New World primates can. They are the smallest South American monkeys and have clawlike nails on all fingers and toes except the big toe, which has a flat nail. These natural crampons, together with their specialized teeth, enable them to cling to vertical trunks while gouging the bark to feed on the gum that lies beneath.

Different teeth

Marmosets and tamarins have fewer teeth than their New World counterparts: 32, the same as Old World monkeys. Dentition also differs between these two species. Tamarins have longer canine teeth that protrude above the incisors, and marmosets have larger incisors, reflecting their greater dependence on gum when other types of foods are scarce. Marmosets use their large, sharp incisors to extract gum from trees. They sink their teeth into the bark and then lick the sticky liquid that oozes out. Although it is rich in energy, the gum is difficult to digest; special bacteria in the cecum (a chamber in the gut) help to break it down.

Since marmosets are able to rely on these sticky substances when the seasonal fruit supply runs low, they can live in places where other food sources are scarce, such as savanna, and in habitats greatly disturbed by humans. In fact, some marmosets live in parks and gardens. By contrast, tamarins eat more fruit and less gum than marmosets, being able to access only the gum that is already seeping from trees. When the fruit supply dwindles, they often revert to nectar as an alternative.

Almost all marmoset and tamarin species also hunt for animal protein. Large insects such as grasshoppers and cockroaches, juicy grubs and even small vertebrates such as lizards and frogs are captured. When hunting, these monkeys wait on a branch, intently scanning the surrounding area for signs of movement. They move in a characteristically jerky manner, making small, quick movements and cocking their heads from side to side. If one of them spots a suitable meal, it quickly pounces, seizing the unsuspecting insect with both hands and quickly devouring it. Grubs may also be foraged from within tree crevices, and vertebrates may be unearthed from leaf litter on the forest floor. Some species of marmosets have even been observed following processions of army ants to catch insects flushed from their hiding places by the ants.

Social structure

Group sizes vary greatly in marmosets and tamarins but average between six and eight members. Marmosets tend to form larger parties—the Black-crowned Dwarf Marmoset, for example, has been seen in groups as large as 30. Some groups, however, consist of only a single breeding pair and its offspring. Group members often groom each other and like to sleep close together in tree holes or tangles of vines. Fights usually break out only over food. When other groups of the same species are encountered, aggression is often expressed with visual displays. Some species stand on their hind legs and stare at their opponents.

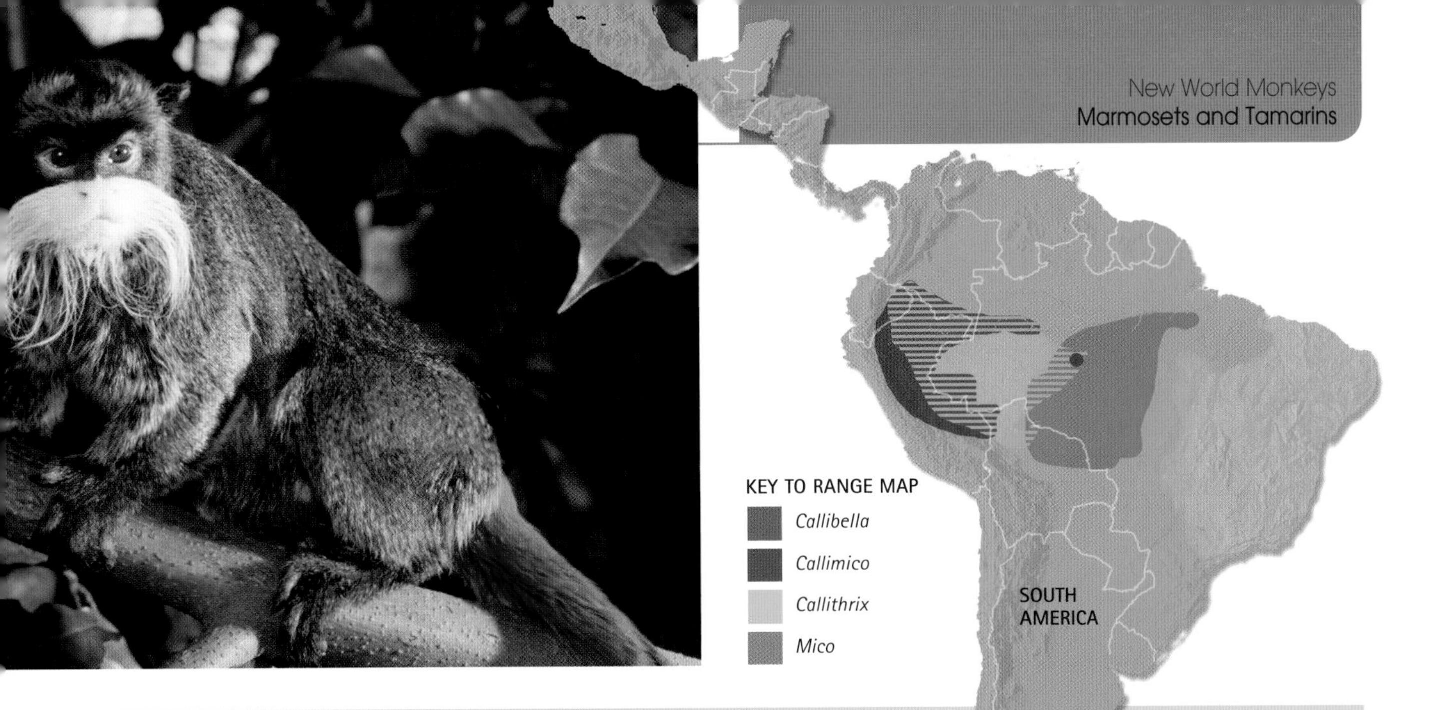

MAIN MARMOSET SPECIES

Scientific name	Common name	Where they live	Red List	CITES
Callibella humilis	Black-crowned Dwarf Marmoset	Brazil	LC	II
Callimico goeldii	Goeldi's Monkey	Bolivia, Colombia, Peru✪	NT	I
Callithrix aurita	Buffy Tufted-ear Marmoset	Brazil	EN	I
Callithrix flaviceps	Buffy-headed Marmoset	Brazil	EN	I
Callithrix geoffroyi	Geoffroy's Tufted-ear Marmoset	Brazil	VU	I
Callithrix jacchus	Common Marmoset	Brazil	LC	II
Callithrix kuhlii	Wied's Black Tufted-ear Marmoset	Brazil	LC	II
Callithrix penicillata	Black Tufted-ear Marmoset	Brazil	LC	II
Callithrix pygmaea	Pygmy Marmoset	Brazil✪, Ecuador, Peru✪	LC	II
Mico acariensis	Rio Acari Marmoset	Brazil	LC	II
Mico argentata	Silvery Marmoset	Brazil	LC	II
Mico chrysoleucus	Golden-white Tassel-ear Marmoset	Brazil	DD	II
Mico emiliae	Emilia's, or Snethlage's, Marmoset	Brazil	LC	II
Mico humeralifer	Black and White Tassel-ear Marmoset	Brazil	LC	II
Mico intermedius	Aripuanã Marmoset	Brazil	LC	II
Mico leucippe	Golden-white Bare-eared Marmoset	Brazil	DD	II
Mico manicorensis	Manicoré Marmoset	Brazil	LC	II
Mico marcai	Marca's Marmoset	Brazil	DD	II
Mico mauesi	Maués Marmoset	Brazil	LC	II
Mico melanurus	Black-tailed Marmoset	Brazil	NE	II
Mico nigriceps	Black-headed Marmoset	Brazil	DD	II
Mico saterei	Sateré Marmoset	Brazil	DD	II

RED LIST: CR = Critically Endangered EN = Endangered VU = Vulnerable NT = Near Threatened LC = Least Concern DD = Data Deficient NE = Not Evaluated ✪ = Best place to watch

Others puff themselves up, making the hairs on their body and tail stand on end to give the illusion of larger body size. The Golden-headed Lion Tamarin flares its fiery crest and flicks its tongue in and out if it feels threatened. Scent marking is also used by marmosets and tamarins as a means of communicating, as is vocalizing using a variety of shrill squeaks and clicking noises.

It is not uncommon to see large groups of marmosets and tamarins containing two or more species. These mixed-species troops benefit from each others' company by increasing overall vigilance in spotting predators. They also become more effective predators themselves, because each individual spends less time watching for danger. As in some human families with twins, in marmoset and tamarin groups everyone helps to care for the young. The infants are carried, food is shared and turns are taken at babysitting duty. But while some individuals enjoy the benefits of this cooperation, others are actively prevented from reproducing. This leads to fierce competition for the right to breed, which can have deadly consequences for some.

PRIMATE PROFILE
MARMOSETS AND TAMARINS

SIZE Head and body length: 7–16 in. (18–40 cm); tail: 6–16½ in. (15–42 cm); weight: 4 oz.–1½ lb. (125–625 g)

APPEARANCE Thick, soft fur (color varies between species); tail rings; various manes, plumes and tufts according to species; thumbs are not opposable

HABITAT Lowland, semideciduous dry forest; gallery forest

DIET Fruit, flowers, nectar, sap, insects, small animals

LIFE HISTORY Gestation: 4–5 months; sexual maturity: 2 years; life span: up to 12 years

BEHAVIOR Lives in groups of up to 20; grooming behaviors; active during the day; territorial

Left The Spix's Saddleback Tamarin uses a vertical clinging posture as it searches for animal prey such as crickets, bugs and frogs, of which it may catch and eat up to 10 each day.

Cooperative breeding

Marmosets and tamarins differ from most primates in their reproduction and associated behaviors. All species within this group practice a unique form of cooperative breeding involving almost all members of the group. There may be one, two or more breeding males within a group and usually one or two reproducing females. These formations are rather flexible.

Marmosets and tamarins can produce up to two litters a year. With the exception of Goeldi's Monkey (which, like most other primates, produces single offspring), the majority of marmoset and tamarin births are of nonidentical twins—such regular twinning is unique among primates. The male will stand behind the female as she gives birth, taking the firstborn and placing it on his back while he waits for the next tiny infant to emerge. Over the next few days he and other group members will carry the young on their back, passing them back to their mother to be suckled.

The young mature very quickly, some reaching sexual maturity at around one year old. Females may conceive again as early as two weeks after giving birth. In other primates ovulation is usually suppressed as long as the mother produces milk. Brothers, sisters, cousins and even some unrelated group members may help to care for the young by babysitting, food sharing and even nursing. Caretakers do benefit themselves, though: babysitting is good practice for when they produce young themselves. Moreover, if they are related, they are helping their kin, so this behavior is beneficial in evolutionary terms.

MAIN TAMARIN SPECIES

Scientific name	Common name	Where they live	Red List	CITES
Leontopithecus caissara	Black-faced Lion Tamarin	Brazil	CR	I
Leontopithecus chrysomelas	Golden-headed Lion Tamarin	Brazil	EN	I
Leontopithecus chrysopygus	Black Lion Tamarin	Brazil	CR	I
Leontopithecus rosalia	Golden Lion Tamarin	Brazil	EN	I
Saguinus bicolor	Pied Bare-face Tamarin	Brazil	CR	I
Saguinus fuscicollis	Spix's Saddleback Tamarin	Bolivia, Brazil, Colombia, Ecuador✪, Peru	LC	II
Saguinus geoffroyi	Geoffroy's Tamarin	Colombia, Costa Rica	LC	I
Saguinus graellsi	Graell's Black-mantled Tamarin	Colombia, Ecuador, Peru	NE	II
Saguinus imperator	Emperor Tamarin	Brazil, Peru	LC	II
Saguinus inustus	Mottled-face Tamarin	Brazil, Colombia	LC	II
Saguinus labiatus	Red-bellied Tamarin	Brazil, Peru✪	LC	II
Saguinus leucopus	Silvery-brown Bare-face Tamarin	Colombia	VU	I
Saguinus martinsi	Martin's Bare-face Tamarin	Brazil	LC	II
Saguinus midas	Golden-handed Tamarin	Brazil, French Guiana, Guyana, Suriname	LC	II
Saguinus mystax	Spix's Mustached Tamarin	Brazil, Peru	LC	II
Saguinus niger	Black-handed Tamarin	Brazil	LC	II
Saguinus nigricollis	Spix's Black-mantled Tamarin	Brazil, Ecuador, Peru	LC	II
Saguinus oedipus	Cottontop Tamarin	Colombia, Panama	EN	I
Saguinus tripartitus	Golden-mantled Saddleback Tamarin	Ecuador	LC	II

RED LIST: CR = Critically Endangered EN = Endangered VU = Vulnerable NT = Near Threatened LC = Least Concern DD = Data Deficient NE = Not Evaluated ✪ = Best place to watch

Subordinate females

The dominant female in a group is the only one that breeds, and she has lots of helping hands to bring up the young. A natural consequence of this is that reproduction in other females is actively suppressed—such suppression is common among daughters still living in their family group. Indeed, the company females keep plays a big part in this mechanism: the mere presence of a more dominant female in the group can prevent a subordinate female from ovulating, and in some species females need to be around a nonrelated male to be able to conceive—these effects may be due to the secretion of chemical messengers called pheromones, and this mechanism may help to prevent inbreeding. A dominant female may also prevent subordinates from mating by displaying or even physically intervening to interrupt a courtship.

However, this complex mechanism for providing childcare and restricting opportunities for breeding is not always cooperative. Since only the dominant members in a group are permitted to breed, opportunities to reproduce are few, and this creates fierce competition. A dominant female may even be driven to murder: if she is pregnant and a subordinate female has young, she may kill them so that her offspring will receive better care from the group. This behavior may also play a part in reproductive suppression—a subordinate female may not reproduce because her young have less chance of survival, so producing them at all is a waste of precious energy and resources.

These unusual primate societies provide excellent case studies for scientists studying evolution. With their small size and short generation time, marmosets and tamarins are considered ideal subjects for the testing of models and theories. But such theories may become purely academic if forest destruction continues. Many marmoset and tamarin populations are dwindling—indeed, this group contains some of the most critically endangered primates. Small geographic ranges coupled with high levels of habitat disturbance make many of them particularly vulnerable to extinction.

LEFT The large eyes and nocturnal habits of the Douroucouli, or Northern Night Monkey, are typical of all monkeys in this family.

Night Monkeys

It is after dark in the New World. The only primates awake in the region's forests are the night or owl monkeys—and a few dedicated primatologists. While the monkeys leap skillfully between the trees, the scientists have to stumble through the thick vegetation, trailing radio-tracking equipment and night-vision binoculars. Despite these difficulties, a handful of researchers have stuck it out in the dark to learn more about the world's only nocturnal monkeys.

Night monkeys are rather peculiar primates. Also called owl monkeys, they are the only monkey species to specialize in foraging in the dark. Moreover, they are also among the few monogamous primates, and it is the father that looks after the young. With the exception of nocturnal prosimians, primates are not well adapted for life at night: they rely mainly on vision for many aspects of their lives, such as finding food and moving around. Night monkeys, however, have evolved features to equip them for a successful night life. Their huge eyes are specially adapted to capture as much light as possible, and although they can't see much color, this trade-off gives them better vision in low light levels.

Night monkeys have not always been nocturnal—they evolved from an ancestor that was active during the day. They may have chosen this nocturnal lifestyle to avoid predators or competition with other primates. Attempts to explain this evolution are complicated, however, by the Southern Night Monkey. This species is cathemeral—it is active at varying times throughout the day and night. Although it shows a preference for nocturnal activity, long-term research has revealed that it spends more time awake during the day if the night is particularly cold or dark.

SOUTH AMERICA

KEY TO RANGE MAP
Aotus nigriceps
Aotus trivirgatus

Until 1983, night monkeys were thought to consist of a single species that was widespread across Central America and northern South America. Then it emerged that there were in fact two groups—gray-necked monkeys in the north and red-necked monkeys in the south, differing both in their coats and genetics. The total number of species is yet to be resolved; it is thought that up to 11 species may exist, but until further genetic studies have been completed scientists cannot be sure.

MAIN NIGHT MONKEY SPECIES

Scientific name	Common name	Where they live	Red List	CITES
Aotus nigriceps	Black-headed or Southern (Red-necked) Night (or Owl) Monkey	Brazil, Peru✪	LC	II
Aotus trivirgatus	Douroucouli or Northern (Gray-necked) Night (or Owl) Monkey	Brazil, Venezuela	LC	II

RED LIST: CR = Critically Endangered EN = Endangered VU = Vulnerable NT = Near Threatened LC = Least Concern DD = Data Deficient NE = Not Evaluated ✪ = Best place to watch

Family groups

Night monkeys live in family groups of between two and six members, consisting of a male and a female and their offspring. It had long been assumed that these monogamous families were stable and long-lasting, but new studies have revealed a frequent turnover of adults in the pairs, so the social system is now described as "serial monogamy." New pairs are formed as, following a few days of aggressive interactions, newcomers displace a partner of the same sex.

When the single infant is born, the mother carries it for the first week, but then the male takes over completely. He passes the youngster to its mother to be suckled, but as soon as the infant has had its fill, she bites it until it returns to the male. He carries out every aspect of parenting, including carrying, playing, grooming and sharing food once the juvenile has been weaned. This high level of paternal care is seen in no other primates with the exception of titi monkeys (see page 84). Young night monkeys generally leave their parents at around three years of age and begin a solitary journey in search of a mate.

PRIMATE PROFILE
NIGHT MONKEYS

SIZE Head and body length: 9½–14½ in. (24–37 cm); weight: 1¼–2¼ lb. (600–1,000 g)

APPEARANCE Coat color variable—usually gray above and pale below; black-and-white markings around eyes; large eyes; some groups have red neck coloration, others have gray necks

HABITAT Rainforest; dry forest; montane forest

DIET Fruit, leaves, flowers, sap, insects, small animals

LIFE HISTORY Gestation: about 4 months; sexual maturity: not known; life span: up to 20 years

BEHAVIOR Active at night; monogamous; lives in small family groups; marks territories with urine and scent; makes owl-like hooting calls

Feeding and communicating

Despite living in dark forests, night monkeys are, impressively, able to snatch insects out of the air as they fly past, or they grab them from branches. However, fruit makes up the main element of their diet. At around sundown, each family of night monkeys wakes up and emerges from the tree hole or thicket that the group shares during the day. The monkeys do not travel far during the night, but go further on brighter nights, and up to twice as far under a full moon. Night monkeys are territorial, and if other groups are encountered, severe fights may break out. The monkeys run on all fours along branches, leaping across gaps.

Group members maintain contact with each other by uttering sounds rather like belches, hoots and whoops. In common with some prosimians, night monkeys practice "urine washing," covering their hands and feet in urine so that their scent is laid down on the branches that they walk along. When two night monkeys meet for the first time, they will spend a long time curiously sniffing each other. There are still many gaps in our knowledge of night monkeys, but continuing research promises more answers in the future.

Left Night monkeys not only look like owls, but hoot like them too, uttering low-frequency calls that travel far through the forest. A captive Southern Night Monkey from Peru is seen here.

Facing page Dusky Titi Monkeys form monogamous pairs. When producing milk, the mother leads the family while the male carries the infant.

Titi Monkeys

Titi monkeys form devoted pairs that are very close emotionally, and they become extremely distressed when separated. They remain close at all times and spend a lot of time grooming each other, which further strengthens their bond. Other signs of affection include intertwining their tails, holding hands, huddling, nuzzling, lip-smacking and gently cuddling.

Titi monkeys were once thought to make up only a moderately diverse group, but a number of recent taxonomic revisions suggest that there may be anything up to 30 species in their genus (*Callicebus*). This makes them second among the New World monkeys in terms of number of species (marmosets and tamarins top the diversity chart, with 41 species currently recognized).

Titi monkeys embody a rare phenomenon, being monogamous primates that mate for life. They live in dense vegetation in the Amazonian and Atlantic forests of South America. Breeding pairs are accompanied by offspring, so groups may grow to include as many as seven members. Youngsters leave the family at two to four years old and may even be allowed a piece of their parents' territory to start them off in life.

Above The Masked Titi Monkey has a black forehead and sideburns. Its tail has a thick coat of fur and is not prehensile (able to grasp).

The first infant is born about a year after a new pair is established, and subsequent births are usually at one-year intervals. During the first week of an infant's life, the mother carries it around for only 20 percent of the time, with the father taking over at other times. Later he becomes the youngster's chief caretaker, carrying it at all times except when it needs suckling. The mother refuses to hold the baby any longer than necessary, and when it has finished feeding, she will nip it or rub it against a branch until the father takes over again. The reasons for this unusually high level of paternal care (seen only in titi and night monkeys within primates) are not fully understood, but some scientists have suggested that it may be related to the high energy demands that producing milk places on the female. Also, a male may be more willing to put effort into infant-rearing in a monogamous relationship, because he can be more certain that the infant belongs to him and not one of his rivals.

PRIMATE PROFILE
TITI MONKEYS

SIZE Head and body length: 9½–24 in. (24–61 cm); weight: 1–4½ lb. (0.5–2 kg)

APPEARANCE Thick fur of variable coloration; often with contrasting face and tail markings

HABITAT Wide range of forest habitats, including rainforest, gallery forest, coastal forest, montane forest and flooded forest

DIET Fruit, seeds, leaves, insects, small vertebrates

LIFE HISTORY Gestation: about 5 months; sexual maturity 24–36 months; life span about 12 years

BEHAVIOR Active in the day; tree-living; usually monogamous; lives in family groups of 2–7 individuals, usually led by the male parent; reinforces social bonds by grooming; birdlike vocalizations

Life at lower levels

Titi monkeys live in trees, spending most of their time in the lower levels of the forest, although they occasionally visit the canopy at the top. They do not come to the ground often, but when they do, they bound along the floor, reaching heights of up to 3 ft. (1 m) with each leap. When moving through the trees, they tend to clamber and leap using all four limbs.

At dawn, a titi group will emerge from the tangled vines in which the monkeys have rested together. The first activity on the family's agenda is to head to the edge of their territory to reinforce the boundaries and see if any neighboring groups are nearby. Loud, distinctive vocalizations reverberate around the trees during these shows of territoriality, and these can be heard throughout the forest. The adult male begins the call, and if it is

ranges between ¼ and 3 miles (0.4 and 5 km) in a day, and its territory typically covers up to 2 square miles (5.5 sq km). As dusk approaches, the group heads back toward a sleeping site to settle down for the night.

All titi species are considered vulnerable, being threatened by hunting and loss of habitat for road building and agriculture. It is vital to preserve the habitats in which they live, to protect their diversity of species and the huge variety of other wildlife that makes up the forest ecosystem.

answered by a nearby group, the adult female joins in to perform a resonant duet. When these vocal exchanges between groups are over, a final retort is uttered—a characteristic "gobbling," promptly returned by the neighbors.

Fruit eaters

After these elaborate displays comes breakfast. Titi monkeys are primarily fruit eaters, but they also eat leaves, and some species include invertebrates in their diet. They are able to cling onto vertical trunks when feeding, using their fingers and toes to hold on. Titis catch insects such as ants, spiders and butterflies by snatching them from branches or even out of the air. Feeding normally takes place in two sessions, one in the morning and the other later in the afternoon, after a midday rest. A family of titis

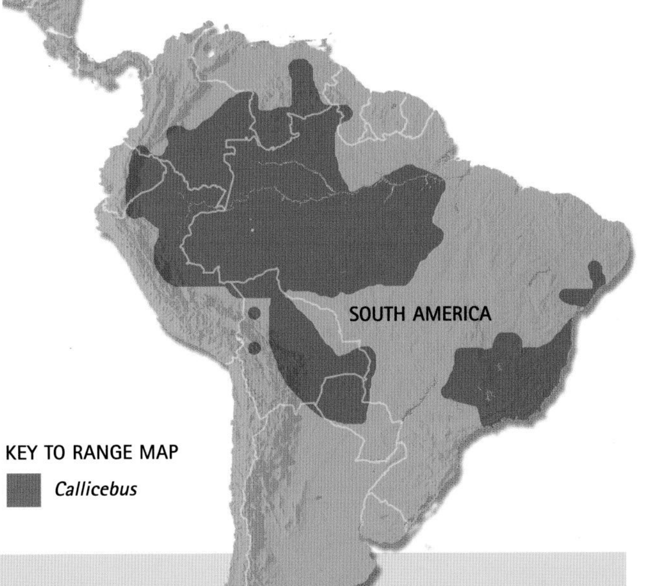

MAIN TITI MONKEY SPECIES

Scientific name	Common name	Where they live	Red List	CITES
Callicebus brunneus	Brown Titi Monkey	Brazil, Peru	LC	II
Callicebus caligatus	Chestnut-bellied Titi Monkey	Brazil, Peru	LC	II
Callicebus cinerascens	Ashy Titi Monkey	Brazil	LC	II
Callicebus cupreus	Red Titi Monkey	Brazil, Peru	LC	II
Callicebus donacophilus	Bolivian Gray Titi Monkey	Bolivia	LC	II
Callicebus dubius	Hershkovitz' Titi Monkey	Brazil	LC	II
Callicebus hoffmannsi	Hoffmann's Titi Monkey	Brazil	LC	II
Callicebus modestus	Rio Beni Titi Monkey	Bolivia	VU	II
Callicebus moloch	Dusky Titi Monkey	Brazil	LC	II
Callicebus oenanthe	Andean Titi Monkey	Peru	VU	II
Callicebus olallae	Beni Titi Monkey	Peru	VU	II
Callicebus personatus	Masked Titi Monkey	Brazil✪	VU	II
Callicebus torquatus	Collared Titi or Widow Monkey	Brazil, Colombia, Venezuela	LC	II

RED LIST: CR = Critically Endangered EN = Endangered VU = Vulnerable NT = Near Threatened LC = Least Concern DD = Data Deficient NE = Not Evaluated ✪ = Best place to watch

Capuchins and Squirrel Monkeys

Capuchin monkeys are now known to be skilled tool-users, as a clearer picture emerges of these highly intelligent and sociable primates. Squirrel monkeys do not use tools but, like capuchins, they do form diverse and complex societies. These two groups make up the subfamily Cebidae and are among the most familiar of monkeys; capuchins were once popular with organ-grinders, and, sadly, many are still captured for the illegal pet trade.

Scientists have never really agreed exactly where in the primate family tree capuchins and squirrel monkeys should be placed. There is an ongoing debate surrounding their relationship to each other and to other families such as night monkeys (see pages 82–83) and the marmosets and tamarins (see pages 78–81). Five species of capuchins are recognized, together with 29 subspecies, several of which may eventually be given full species status. The current consensus on squirrel monkeys is that there are 5 species and about 12 subspecies.

Dextrous and resourceful

Capuchins and squirrel monkeys are the only New World monkeys that have thumbs that are opposable—although in a more limited sense, known as pseudo-opposable, than the fully opposable thumbs of Old World monkeys and apes. (Prosimians also have pseudo-opposable thumbs.) Combined with their independent finger movement, this makes capuchins the most dextrous monkeys in the New World and the only ones able to pick up small objects using their fingers independently—squirrel monkeys are less dextrous.

This dexterity, combined with a large brain-to-body size ratio, has enabled capuchins to become incredibly resourceful. They forage for a very wide range of foods, many of which are avoided by other primates due to the difficulty involved in accessing them. Capuchins have the thickest tooth enamel of any nonhuman primate, which helps them to crack open hard seeds and nuts, and they are adept at coming up with clever ways to overcome the defenses put up by their food: hard-shelled nuts are smashed against rocks or branches, or sometimes smashed with rocks; spines or hairs are removed from caterpillars or plants by rubbing them against a branch, using leaves to protect the hands; small branches are used to probe into crevices; and large branches may even be used to club a venomous snake.

Fruit makes up the majority of the capuchin diet. It also forms the bulk of the squirrel monkeys' diet, but they eat more insects than capuchins. Small vertebrates, such as the colorful tree frogs that bejewel the Amazon forests, may also be captured and eaten. Flowers, nectar and leaves are on the menu as well.

Right The Brown-capped Capuchin has astonished scientists by using tools both in the wild and, as here in captivity, in ways similar to those of Chimpanzees. This individual is smashing a nut against a stone to crack its shell.

PRIMATE PROFILE
CAPUCHINS

SIZE Head and body length: 12–22½ in. (31–57 cm); weight: 2¼–9¼ lb. (1.1–4.3 kg); males larger than females

APPEARANCE Coat color very variable between species; semiprehensile tail, with a distinctive coil at the tip

HABITAT Primary deciduous, mangrove, gallery, lowland and montane forest; secondary rainforest

DIET Fruit, seeds, leaves, gums, flowers, birds' eggs, insects, small vertebrates, crabs, oysters

LIFE HISTORY Gestation: up to 6 months; sexual maturity: male 4 years, female 7 years; life span: up to 40 years

BEHAVIOR Active in the day; tree-living; lives in groups of 8–33 members; communicates by vocalization and display; frequently uses tools

Above A Brown-capped Capuchin reaches for fruit at the tips of branches in a tree in Manu National Park, Peru, demonstrating the usefulness of a prehensile (grasping) tail when feeding in flimsy branches; a fifth limb gives an extra safety anchor in case one of the twigs being gripped for support breaks.

Tagging along

As they move through the trees, groups of capuchins are often accompanied by squirrel monkeys. The latter are smaller and an easier target for predators, especially during the breeding season when vulnerable infants are around, but they reduce the risk by often foraging alongside capuchins to take advantage of their greater predator-spotting vigilance. Capuchins may also drop tasty food morsels that squirrel monkeys cannot normally obtain, such as hard-shelled nuts. These tidbits are scavenged from the forest floor, and capuchins usually tolerate their smaller relatives.

Squirrel monkeys and capuchins are found in virtually all forest types across South and Central America. Much of their day is spent up in the canopy, although capuchins visit the ground more often. They run, clamber and leap through the trees using all four limbs, but squirrel monkeys leap more often, using their tail for balance (their tails are prehensile, or able to grasp, in infants only). Capuchin tails are more prehensile and are used for balance and support. Like most monkeys, capuchins and squirrel monkeys spend most of their day feeding and most of the night sleeping. Squirrel monkeys sleep high up in the canopy; capuchins sleep in tall trees.

Social structure

The number of squirrel monkeys in a group varies from about 15 to 75. Groups of up to 300 have sometimes been seen, but these are thought to be the coming together of several smaller groups. There is huge variation in squirrel monkey social systems. Some are egalitarian (all individuals are equal), some are male-dominant, and in one species—the Bolivian Squirrel Monkey—females are dominant. This species is one of only two tropical New World primates to include all-male bachelor groups.

Capuchins typically form groups of 16 to 20. When males mature, they tend to leave the group, but females remain with their families. Males are usually individually dominant over females, but a coalition of two or more determined females can oust even the alpha male from a good feeding spot.

In squirrel monkeys, communication is by sound, sight and scent. Twenty different vocalizations have been noted. Visual displays are also used to communicate things such as dominance—for example, a male may present his penis or urinate on a subordinate, in order to assert or maintain his dominance.

Capuchin monkeys have complex and varied methods of communicating. Their vocal repertoire includes an intricate alarm system, with different calls for different predator types. Urine washing and chest rubbing are also important aspects of communicating by scent. In addition, capuchins use a variety of visual displays. If two males have been separated for a long time, they will run toward each other and embrace, vocalizing loudly. Some groups have been seen banging stones together to warn each other of approaching predators—a fascinating example of tool-use that has come to light only recently.

Reproduction

Squirrel monkeys have one of the most unusual reproduction systems seen in primates. In the months leading up to the birth season, male Red-backed and Bolivian Squirrel Monkeys undergo drastic body weight increases of up to 22 percent. This extra weight is thought to come from retention of water rather than accumulation of fat, and it is concentrated around the head, shoulders, arms and ribs. Although there is not normally much difference in size between males of these species, some are able to gain more weight than others—and the females seem to prefer these larger males.

Around the time of the breeding season, groups of up to 16 males will routinely chase a female and pin her down, inspecting her for signs of receptivity. This mobbing may occur as often as twice an hour at the peak of the season. Female squirrel monkeys synchronize the birth of their infants, which reduces the chances of them falling prey to predators. The mother provides almost all of the parental care, carrying the single infant on her back for the first month of its life.

By contrast, capuchins do not have a strict breeding season but do synchronize births with times of plentiful food. They also give birth to single infants, and for the first few weeks the young capuchin is carried by its mother. It then spends anything up to 70 percent of its time with other group members, including the males, who help with its rearing.

Courtship behavior

Capuchin monkeys have some rather interesting sexual traits. Sex is usually initiated by females, who go to some lengths to catch the eye of a suitable partner (usually the alpha male). A female may begin with a suggestive raising of the eyebrows or pouting of the lips. While edging closer to the male, she cocks her head from side to side, stroking her chest and making gentle noises. She may then strike a pose, looking back at him over her shoulder, or sit facing him spreading her thighs. If the male still shows no interest at this point (which is unusual), she may touch him and quickly run away. When he eventually gives in to her persuasions, the male will mirror her actions, grinning and vocalizing. A salsa-style dance of forward and backward moves then ensues, complete with 180° pirouettes. The pair of capuchins may look between their legs or over their shoulders to maintain their fixed gaze on each other. After they have finally mated, they may continue this bizarre performance.

PRIMATE PROFILE
SQUIRREL MONKEYS

SIZE Head and body length: 10–14 in. (25–36 cm); weight: 1¾–2¼ lb. (750–1,100 g)

APPEARANCE Short, thick fur; mainly white face; gray top of head; large ears; short, but well-developed thumb

HABITAT Mature, moist upland forest; mangroves

DIET Fruit, seeds, leaves, gums, flowers, insects and spiders, small vertebrates

LIFE HISTORY Gestation: 5–6 months; sexual maturity 2 years; life span: 21 years

BEHAVIOR Active in the day; tree-living in groups of up to 550 males and females

Above The highly endangered Red-backed Squirrel Monkey has a body length of about 12 in. (30 cm). It moves by leaping from one branch to another rather than clambering along branches.

The production of offspring is not the only reason why capuchins mate. Sexual behavior between males may alleviate tension within the group after fighting or the arrival of newcomers and also helps to form coalitions. This use of sexual behavior for social purposes is comparable to that seen in Bonobos in Africa (see pages 162–165).

Although capuchins are some of the most studied New World tropical primates, there is still much less known about them than about the monkeys and apes of the Old World. More long-term field studies are needed to fully understand the complex range of behavior, social structures and reproductive systems in both capuchins and squirrel monkeys.

When tool-use was first reported in wild capuchins (see pages 20–21), it was greeted with some skepticism. However, in 2004, a multinational research consortium was set up to investigate it further. The results may lead to new insights into the evolution of human behavior, giving yet another reason to protect these primates and their habitats. Some capuchin and squirrel monkey species are vulnerable or endangered, and as habitat destruction continues in their Amazonian forest home, the risk of losing these remarkable monkeys is increasing.

CAPUCHIN AND SQUIRREL MONKEY SPECIES

Scientific name	Common name	Where they live	Red List	CITES
Cebus albifrons	White-fronted Capuchin	Brazil, Colombia, Ecuador, Peru	LC	II
Cebus apella	Tufted, Black-capped, or Brown-capped Capuchin	Northern and central South America from Colombia and Ecuador to coastal Brazil (French Guiana✪, Peru✪)	LC	II
Cebus capucinus	White-faced, White-headed, or White-throated Capuchin	Honduras to northwestern Ecuador	LC	II
Cebus kaapori	Ka'apor Capuchin	Brazil	VU	II
Cebus olivaceus	Weeper or Wedge-capped Capuchin	Brazil, French Guiana, Guyana, Suriname, Venezuela✪	LC	II
Saimiri boliviensis	Bolivian or Black-capped Squirrel Monkey	Bolivia, Brazil, Peru	LC	II
Saimiri oerstedii	Red-backed or Central American Squirrel Monkey	Costa Rica✪ to the Pacific coast of Panama	EN	I
Saimiri sciureus	Common or South American Squirrel Monkey	Brazil, Colombia, Ecuador, French Guiana, Guyana, Suriname, Venezuela	LC	II
Saimiri ustus	Golden-backed or Bare-eared Squirrel Monkey	Northwestern Brazil	LC	II
Saimiri vanzolinii	Black Squirrel Monkey	Northwestern Brazil	VU	II

RED LIST: CR = Critically Endangered EN = Endangered VU = Vulnerable NT = Near Threatened LC = Least Concern DD = Data Deficient NE = Not Evaluated ✪ = Best place to watch

Sakis and Uakaris

Sakis and uakaris are very peculiar-looking primates. Uakaris have a tiny, bald, bright red head poking out from a huge, shaggy orange, white or black coat. White-faced Sakis have an equally odd-looking face that contrasts sharply against their dark body hair. Bearded sakis, on the other hand, go to the opposite extreme, sporting a spectacular bouffant hairstyle and characteristic beard, which is present even in females.

Above The skin of the face of the Red Uakari is gray in juveniles and becomes red only with maturity, as in the adult pictured here. The young also have gray hair on the crown, which is lost as they mature.

Sakis and uakaris (pronounced "wakaris") make up the Pitheciine subfamily (some authors put them with titi monkeys in the family Pitheciidae, *see* pages 84–85). Their taxonomy, like that of most primates, has recently been revised. Seven species of beardless saki, with a few subspecies, and two species of bearded saki are currently recognized; recent studies suggest promoting at least three bearded saki subspecies to full species, which would bring the total number of sakis to 10 or more species. Until 2008, three species of uakaris were recognized: the Black-headed, Bare-headed and Red Uakari. Then a fourth species, Ayres' Black Uakari, was discovered in montane rainforest in western Brazil, an unexpected habitat for uakaris. Bearded sakis and uakaris are similar in many respects; the five species of *Pithecia*, smaller beardless sakis, share similarities with both their bearded relatives and the titi monkeys.

Unripe food

Many fruit-eating primates tend to eat soft, fleshy fruits and avoid unripe fruits, nuts and seeds because their hard exterior makes it difficult to break them open. Sakis and uakaris have taken advantage of the lack of competition for such foods, evolving specialized canines and strong jaw muscles to break through the tough outer cases. Their long incisors also project forward, enabling them to scrape out the edible interior. Fruit is the most important food for these primates, and in the case of White-faced Sakis, two fruit species account for half of their entire fruit intake. However, seeds may also contribute up to a third of the diet of sakis and uakaris, providing a reliable source of energy when fruit availability is low. Beardless sakis seem to have a slightly broader diet that contains more leaves, reflected in their dental structures, which are less specialized for eating hard fruits and seeds.

Sakis and uakaris supplement their tough diet with insects, birds' eggs and small vertebrates such as bats, which are found sleeping in tree holes during the day and are enthusiastically torn apart and devoured. These monkeys will occasionally attack wasp and termite nests, which are high in iron, and they also eat soil, which may provide extra nutrients. Lacking an opposable thumb, sakis and uakaris hold their food between their index and middle fingers. At feeding times, groups travel very quickly through the trees, stopping for intense feeding bouts then swiftly moving on—so quickly in fact, that if a group member does not keep up, it may be left behind. Such stragglers often accompany groups of capuchins or squirrel monkeys for a while, until they can rejoin their own species.

Crashing through the forest

Sakis and uakaris make for a magnificent sight when traveling through the trees. The observer is alerted to their arrival by a crashing of branches overhead. The monkeys clamber noisily through the branches and make huge leaps between tree crowns with arms flailing in the air, crashing into the next tree. Uakaris have been seen rocking branches backward and forward in order to build enough momentum to catapult them to adjacent trees and make leaps of up to 20 ft. (6 m). Beardless sakis differ from

PRIMATE PROFILE
SAKIS

SIZE Head and body length: 12–28 in. (30–70 cm); weight: 2¼–8¾ lb. (1–4 kg)

APPEARANCE Thick, coarse fur, growing forward from the nape of the neck; bushy tail

HABITAT Lowland and riverine forest; coastal forest; savanna

DIET Fruit, seeds, leaves, flowers, honey, insects, small animals

LIFE HISTORY Gestation: 5–6 months; sexual maturity: 2–3 years; life span: up to 14 years

BEHAVIOR Active in day; tree-living; variable social groups

Right The prominent white facial hair of the male White-faced Saki is clearly seen here. The females are brown and lack the striking white face, having instead an elegant pale stripe on either side of the nose.

uakaris and bearded sakis in the ways in which they traverse the forest. As well as climbing and running over branches on all fours, they use the vertical clinging and leaping method of locomotion, using their long tails as rudders to steer while in the air.

Social characteristics

Beardless sakis also differ from bearded sakis and uakaris in their social tendencies. Although it seems that a complete picture of saki social structures is yet to be revealed, beardless sakis are thought to form monogamous pairs. They have been seen in various group sizes and compositions, which may be affected by food availability, habitat and the presence of other primates. Small groups containing an adult pair and their offspring have been recorded, as have groups containing a few adults, with up to 12 members. Aggression observed between neighboring groups suggests that beardless sakis may also be territorial.

Bearded sakis and uakaris, by contrast, seem to form larger, more loosely structured groups. Congregations may have more than 100 animals, with smaller groups breaking off to feed. These spectacular gatherings may last for a few days. This Chimpanzee-like "fission-fusion" type of social structure may be influenced by the seasons, food availability and other factors.

Communication between individuals and groups takes many forms. Beardless saki pairs sing duets in much the same way as titi monkeys, to maintain territorial boundaries and strengthen the bond between the monogamous pair. Sakis also produce a distinctive, high-pitched "*squeeeeeeechuck*" call and a "*hoo-hoo*" call to communicate with their groups. Scent is thought to play an important part of social life in all New World monkeys, but the finer details of its functions and uses in sakis and uakaris are yet to be determined. These monkeys are known to mark branches with their scent glands (although bearded sakis are an exception to this behavior), and they exchange scent through bodily contact, by hugging each other. Bearded sakis and uakaris wave their bushy tails in various situations and also shake branches in visual display when angry. Uakaris blush crimson when disturbed and make a loud noise that sounds rather like human laughter.

Raising young

In monogamous primates such as night and titi monkeys, the adult male typically plays an active role in rearing offspring, sometimes becoming the sole carer. Beardless sakis, however, are an exception to this rule, in that the female provides almost all of the care for the single infant. She carries it for the first eight weeks of its life, and then other females in the group may lend a hand. Some males do babysit while the mother forages, but this is low-investment parental care in comparison to that provided by monogamous fathers in other species.

PRIMATE PROFILE
UAKARIS

SIZE Head and body length: 12–22½ in. (30–57 cm); weight: 6–8 lb. (2.7–3.7 kg)

APPEARANCE Long, white, red or black fur (according to species); short tail

HABITAT Swamp and dry-floored forest; mountain forest

DIET Fruit, seeds, leaves, insects, small animals

LIFE HISTORY Gestation: 6 months; sexual maturity: female 4.5 years, male 5.5 years; life span: 20 years

BEHAVIOR Active during day; tree-living; lives in groups of 5–30 individuals; communicates using a variety of calls and facial expressions

Facing page The Black-headed Uakari looks in trees for fruit, seeds and small creatures. Here, an individual clambers through the forest in Manaus, Brazil.

In contrast to their beardless relatives, bearded saki and uakari mothers carry their infants for about nine months, more than twice as long. The slower development of the infant may be due to their larger body size, and this feature may also explain the longer period between births, which is two years compared to 15 months in beardless sakis. As is the case with beardless sakis, the female is the primary care giver.

Monkey mysteries

There is still a long way to go before we can fully appreciate the complexities of saki and uakari societies. To date, relatively few studies have been devoted to these monkeys' social habits, mostly due to their remoteness and lifestyle. Groups are far-ranging and fast-moving, making it very difficult to study them. Much of the uakaris' range is flooded for nine months of the year, and the floodwaters reach depths of up to 23 ft. (7 m), so that often the only means of accessing these areas is by canoe.

Although the impenetrability of much of their range deters researchers from following sakis and uakaris, hunters of these species have proved to be adept at navigating the Amazon tributaries and flooded forests. In addition, habitat clearance on a huge scale continues to cut deep swathes through the Amazonian forests, threatening all species in its wake. Fortunately, in 2006 it was announced that the largest tract of tropical forest in the world—at about 50,000 square miles (140,000 sq km) it has a similar area to the American state of Illinois—is to be set aside as protected reserve. This area crosses international borders and contains more than a quarter of the world's remaining humid tropical forest and thankfully encompasses the range of some of these amazing primate species. If the good news of the announcement is backed up by the introduction of measures to ensure active protection on the ground, this may prove to be a significant step in the right direction toward preserving these still poorly understood primates.

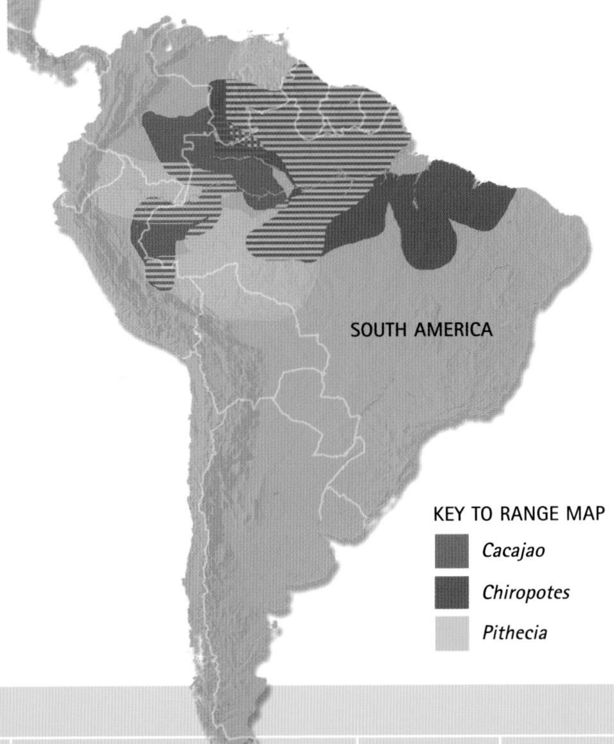

UAKARI AND MAIN SAKI SPECIES

Scientific name	Common name	Where they live	Red List	CITES
Cacajao ayresii	Ayres' Black Uakari	Brazil	NE	I
Cacajao calvus	Bald Uakari	Brazil✪, Peru	NT	I
Cacajao melanocephalus	Black-headed Uakari	Brazil✪, Venezuela	LC	I
Cacajao rubicundus	Red Uakari	Brazil✪, Colombia (?)	VU	I
Chiropotes albinasus	White-nosed Bearded Saki	Brazil	LC	I
Chiropotes satanas	Bearded Saki	Brazil, French Guiana, Guyana, Suriname, Venezuela	EN	II
Pithecia aequatorialis	Equatorial Saki	Ecuador, Peru	LC	II
Pithecia albicans	Buffy Saki	Brazil	LC	II
Pithecia irrorata	Bald-faced Saki	Bolivia, Brazil	LC	II
Pithecia monachus	Monk Saki	Brazil, Colombia, Ecuador, Peru	LC	II
Pithecia pithecia	White-faced Saki	Brazil, French Guiana✪, Guyana, Suriname, Venezuela✪	LC	II

RED LIST: CR = Critically Endangered EN = Endangered VU = Vulnerable NT = Near Threatened LC = Least Concern DD = Data Deficient NE = Not Evaluated ✪ = Best place to watch

Facing page Living high in the canopy, you can catch the early morning sun. Here a young Black-and-gold Howler snuggles up to its mother.

Howler Monkeys

As the sun rises over the tropical forests of the New World, a powerful bellow echoes through the treetops. Louder than any dawn chorus sung by birds, this is how howler monkeys greet the day. Their supersized voice box amplifies the sound to carry up to 3 miles (5 km), informing other groups that this patch of forest is occupied.

Perhaps it was the distinctive call of howler monkeys that led the Maya to believe that they were divine creatures. The ancient Maya civilization worshipped howler monkey gods, and although this culture has long since disappeared, the monkeys still roam the ruins of Mayan cities in Honduras and Guatemala. More of a roaring, guttural whoop than a howl, their powerful call is the loudest sound made by any land animal. The "long calls," as they are known, may serve to maintain distance between groups, avoiding the need for confrontation or aggressive encounters, and thus conserving energy.

Why do howler monkeys need to conserve energy? The answer can be found in the food they eat: a large part of their diet is made up of leaves. Leaves are a poor energy source, so a howler monkey has dental adaptations to grind the leaves better. In addition, large salivary glands produce saliva to break down the poisonous tannins in the leaves. The food then stays in the gut for a long time, giving gut bacteria more time to ferment the foliage. This produces a gaseous by-product, which is reabsorbed and used. In this way, a howler monkey gains the maximum amount of energy from each leaf. Studies of red howler monkeys show that they eat 195 plant species and on average consume 2.7 lb. (1.2 kg) of fresh food per day.

Sedentary lifestyle

Howler monkeys spend up to 80 percent of their time resting so as not to waste energy, further conserving energy by howling a warning rather than patrolling their territory on the lookout for encroachers. After their nap, they may move to a new area to find more food, but they won't move very far—they don't need to, because leaves are always nearby in a forest. The distance they travel in a day amounts to no more than about 765 yards (700 m). Howlers do not eat only leaves, however—fruit is also an important energy source, and the monkeys are important seed dispersers for many trees. Insects are also eaten.

PRIMATE PROFILE
HOWLER MONKEYS

SIZE Head and body length: 22–36 in. (56–92 cm); weight: 3½ oz.–4½ lb. (4–10 kg)

APPEARANCE Coarse coat, longer in head and shoulder areas, usually brown or black (according to species); prehensile (able to grasp) tail; enlarged throat

HABITAT Wide range of forest types; savanna

DIET Fruit, leaves, flowers, buds

LIFE HISTORY Gestation: 6 months; sexual maturity: male 4.5–5.5 years, female 3.5–4.5 years; life span: up to 25 years

BEHAVIOR Active during the day; tree-living; social structure varies; males compete strongly; loud calls

Left The howler monkey equivalent of a roll call reverberates around the forest, announcing that this family of Venezuelan Red Howlers is awake and active and would prefer that the neighbors keep their distance.

Howler monkeys have a wide range and can be seen and heard in a variety of forest types from northern Argentina to the Yucatan Peninsula of Mexico. They move slowly through the trees, using their powerful prehensile tail and all four limbs to grasp branches, and they may hang from the tail while feeding. The tail is also coiled around a branch to act as an anchor while the monkey sleeps.

Home ranges of howler monkeys vary widely depending on the quality of the habitat. Some groups' ranges do not overlap very much; others overlap almost entirely, but the groups keep their distance by calling. The nine different species also vary in how they organize their social lives, and there are differences within species too. Groups seldom exceed 15 members, often led by one alpha male, although groups with two or three males are also common. Females nearly always outnumber the males and are subordinate to them.

Social mobility

Within howler monkey society, members of either sex may leave their family group upon reaching sexual maturity. Males that disperse in this way can usually fight their way into other groups, but females are normally forced to start their own group. Mantled Howlers are an exception. They form male-female groups of 40; either sex may disperse and females are more likely to join a new group than those of other howler species. Ambitious male howlers will go to great lengths to become the alpha male and father the next generation. This may even involve killing the young of other males when taking over a new group.

Howler monkeys have a relatively fast reproduction rate compared to other members of their subfamily (Atelidae), the spider monkeys, woolly monkeys and muriquis (see pages 96–99), and they can recover more quickly from hunting. Even so, the scale of hunting today is not sustainable. In addition, the habitats in which howler monkeys live are threatened by clearance for roads, electricity networks and other types of development. If better protection is not achieved, extinction looms and mornings in the tropical forests of the New World may never sound the same again.

SOUTH AMERICA

KEY TO RANGE MAP

Alouatta

HOWLER MONKEY SPECIES

Scientific name	Common name	Where they live	Red List	CITES
Alouatta belzebul	Red-handed Howler	Brazil✪	LC	II
Alouatta caraya	Black-and-gold Howler	Argentina, Brazil✪, Paraguay	LC	II
Alouatta guariba	Red-and-black Howler	Argentina, Brazil	NT	II
Alouatta macconnelli	Guyanan Red Howler	Brazil, Guyana	VU	II
Alouatta nigerrima	Amazon Black Howler	Brazil	NE	II
Alouatta palliata	Mantled Howler	Mexico to Equador	LC	I
Alouatta pigra	Guatemalan Black Howler	Belize✪, Guatemala✪, Mexico	EN	I
Alouatta sara	Bolivian Red Howler	Bolivia✪	LC	II
Alouatta seniculus	Venezuelan Red Howler	Northern South America (Venezuela✪)	LC	II

RED LIST: CR = Critically Endangered EN = Endangered VU = Vulnerable NT = Near Threatened LC = Least Concern DD = Data Deficient NE = Not Evaluated ✪ = Best place to watch

Left Dangling from its long tail like a spider on a thread, fruit in hand, this White-fronted Spider Monkey is perfectly adapted to a life in the Amazon treetops.

Spider Monkeys, Muriquis and Woolly Monkeys

Spider monkeys are the unrivaled acrobats of the tropical forests of the New World. Their long, spindly limbs and prehensile tail enable them effortlessly to navigate through the towering treetops, as if they had five hands with which to grip.

Watching a group of spider monkeys frolic in the branches overhead, or better still observing them from a treetop platform or walkway, is a truly exciting experience. Like arboreal trapeze artists, they swing from branch to branch with ease, using their tails as a fifth limb. With legs a-dangle, they often hang from just their tails while feeding, looking like a spider hanging from its thread—hence the name.

Spider monkeys are perfectly adapted for this swinging life in the trees. Their hands and feet have four long fingers or toes and virtually no thumb or big toe, turning them into quick-release gripping hooks that are used in under-branch, hand-over-hand brachiation (*see* pages 150–151). Their slender arms are even longer than their lanky legs and have highly flexible shoulder joints. The strong, muscular tail has a hairless pad on the underside, which is rich in touch receptors, so that it can function like an extra hand. Some spider monkeys have been observed using their tail in other useful ways, such as collecting drinking water from deep tree holes.

Above A Northern Muriqui mother, infant firmly anchored tail-in-tail, pauses with four points of contact on the branches, using her tail as a fifth limb while considering where to reach with her spare hand.

PRIMATE PROFILE
SPIDER MONKEYS AND MURIQUIS

SIZE Head and body length: 15–25 in. (38–64 cm); weight: 13–33 lb. (5.5–15 kg)

APPEARANCE Long coat (spider monkeys), short, thick coat (muriquis); color varies according to species; long, prehensile tail; poorly developed or absent thumb

HABITAT Rainforest; semideciduous forest

DIET Fruit, leaves, flowers, seeds, bark, small animals (spider monkeys)

LIFE HISTORY Gestation: 7–8 months; sexual maturity: 6–11 years; life span: up to 30 years

BEHAVIOR Active in daytime; tree-living; lives in groups of variable social structure; territorial; communicates using a variety of calls

Woolly monkeys get around in much the same way as their spidery cousins, but tend to take things somewhat slower. They make shorter, more deliberate movements and don't hang from their tails as much as spider monkeys. They also usually use the tail and a hand at the same time, in contrast to spider monkeys, which often dangle from a single appendage. The strength of the tail's grip is extraordinary—one captive male woolly monkey was seen to lift a person off the ground while hanging by his tail alone wrapped round a vertical rope!

Disputed taxonomy

The relationship between spider and woolly monkeys has yet to be agreed. Muriquis, the largest New World nonhuman primates, are also referred to as woolly spider monkeys. This name, however, is misleading; although they share many characteristics with spider monkeys (and so often have been grouped with them), recent molecular studies suggest that muriquis are in fact more closely related to woolly monkeys. There are two species of muriqui—northern and southern—which occur in only a few areas of forest on the Atlantic coast of Brazil.

Spider monkeys can be found swinging through many of the tropical and deciduous forests of Central and South America. Seven different species have been described, but some taxonomists recognize only four. Woolly monkeys are also widely distributed and live throughout South American rainforests. Five species of woolly monkey are generally accepted.

Flexible societies

Spider monkeys live in a "fission-fusion" type of community, much like Chimpanzees (see pages 162–165). Each group contains up to 40 monkeys and regularly splits into smaller groups to forage and travel together. This fluid social structure is used as a way of coping with a food supply that is variable. Large groups trying to feed together in these conditions would lead to competition within the group; smaller foraging parties are more effective.

The foraging parties usually comprise around three individuals—often a male, a female and her offspring. Sexes may also roam separately; males patrol territory boundaries, where aggressive encounters may occur with neighboring troops. This may include branch shaking and vocalizing.

Different groups keep in touch by long-calling, a sound produced by males that travels up to 1¼ miles (2 km) through the canopy. Sometimes, when subgroups congregate, there may be some tension between the males. As a way of reducing this tension, spider monkeys greet each other warmly by hugging. After embracing, they will sniff each other's chests and genitals.

Woolly monkeys form slightly larger groups than spider monkeys, containing 18 to 45 individuals. Most of the time these groups travel and feed together, but the composition is flexible. Individuals may go and visit other groups, spending a few days or weeks with them, and smaller groups may sometimes break off to feed separately. When larger groups are spread out through the trees, they use vocalizations to keep in contact with each other over distances of several hundred yards. Woolly monkeys also send olfactory messages by chest-rubbing and anogenital marking, and they greet each other by embracing or "sobbing."

Sobbing occurs when a monkey lifts a forearm to cover its eyes while making an *"ogh-ogh"* noise. Woolly monkeys also love to play, which may involve wrestling and chasing.

Northern Muriquis form large groups that tend to remain together. If a group becomes too big, however, the monkeys split into smaller parties, as do Southern Muriquis, which have a fission-fusion society. Because muriquis eat more leaves than fruit, they can more often stay in a cohesive group to feed. If they do break into smaller feeding groups, these are often larger than those formed by spider monkeys.

Above For the first few months of life, an infant Brown Woolly Monkey never loses contact with its mother for a single moment, unless by accident; its survival depends on gripping her fur with hands and feet, and its tail looped around hers provides added security.

Fruit eaters

Spider monkeys mostly forage for fruit, which can represent up to 70 percent of their diet. They particularly enjoy ripe fruit—sometimes so ripe that it is fermented and the monkeys become drunk. Their play then becomes more boisterous, and their forest acrobatics are likely to result in a drunken monkey or two crashing through the branches after a misjudged swing. Leaves are another important food source for spider monkeys, as well as flowers and insects. Woolly monkeys consume a diet consisting mostly of fruit, relying more on seeds when fruit supplies run low. They also nibble on leaves and flowers and may sometimes catch invertebrates.

After spending the night sleeping in the forest canopy, spider and woolly monkeys untangle themselves and set about looking for food. They may travel long distances over the course of the

PRIMATE PROFILE
WOOLLY MONKEYS

SIZE Head and body length: 20–27 in. (51–69 cm); weight: 9¾–22 lb. (4–10 kg)

APPEARANCE Short, dense fur (brown, black or gray, according to species); round head with prominent brow; flat face; short fingers; non-opposable thumb; thick, prehensile tail

HABITAT Swamp and dry-floored forest; mountain forest

DIET Fruit, seeds, leaves, insects, small animals

LIFE HISTORY Gestation: 7–8 months; sexual maturity: 4–5 years; life span: 26 years

BEHAVIOR Active during day; tree-living; lives in groups of 5–30 individuals; communicates by calls and facial expressions

day to find ripe fruit, remembering the same routes through the forest. After a few hours of foraging, they will settle down for a midday rest, resuming feeding in the afternoon.

Mating and parenting behavior

Spider monkey pairs often head to a secluded area of the forest to mate. This makes it difficult for researchers wanting to study spider monkey mating systems, so there is much left to learn. The monkeys are known to give birth to single infants, and although they do have a birth season, the timing of this varies between species. The young are cared for by their mothers and stay near them until they are about four years old. Each mother carries her own infant, with it riding on her back after about six months. When there is a large gap in the canopy, and the infant is too large to carry, she will use her body to make a bridge between the trees. She holds one tree with her hands and the other with her feet and tail while her young clambers along her gangly body as if on a rope bridge. Woolly monkeys, spider monkeys and muriquis reproduce relatively slowly. The age at which they first reproduce is quite late compared to other primates of a similar body size, and there may be a four-year interval between births. This means that populations find it more difficult to recover if numbers fall for any reason.

Hunting is a prime cause of such a drop in numbers. Populations of these monkeys can withstand traditional hunting with blowpipes by tribes living at low density, but once commercial hunters with firearms arrive, numbers soon fall. These monkeys are also highly susceptible to habitat destruction. Consequently, nearly all of them are threatened to at least some extent, and the Yellow-tailed Woolly Monkey and two spider monkeys are among the top 25 most endangered primates in the world.

MAIN SPIDER MONKEY SPECIES

Scientific name	Common name	Where they live	Red List	CITES
Ateles belzebuth	White-fronted Spider Monkey	Brazil, Colombia, Equador, Peru✪	VU	II
Ateles geoffroy	Geoffroy's or Black-handed Spider Monkey	Mexico✪ to Panama (Costa Rica✪)	LC	I
Ateles hybridus	Brown or Variegated Spider Monkey	Colombia, Venezuela	CR	II
Ateles paniscus	Red-faced Spider Monkey	Brazil, French Guiana, Guyana, Suriname	LC	II

MURIQUI SPECIES

Scientific name	Common name	Where they live	Red List	CITES
Brachyteles arachnoides	Southern Muriqui or Woolly Spider Monkey	Brazil✪	EN	I
Brachyteles hypoxanthus	Northern Muriqui or Woolly Spider Monkey	Brazil✪	CR	II

WOOLLY MONKEY SPECIES

Scientific name	Common name	Where they live	Red List	CITES
Lagothrix cana	Gray or Geoffroy's Woolly Monkey	Bolivia, Brazil, Peru	NT	II
Lagothrix lagotricha	Brown or Humboldt's Woolly Monkey	Brazil, Colombia, Equador, Peru	LC	II
Lagothrix lugens	Colombian Woolly Monkey	Colombia, Venezuela	VU	II
Lagothrix poeppigii	Silvery or Poeppig's Woolly Monkey	Brazil, Equador, Peru	NT	II
Oreonax flavicauda	Yellow-tailed or Hendee's Woolly Monkey	Peru	CR	II

RED LIST: CR = Critically Endangered EN = Endangered VU = Vulnerable NT = Near Threatened LC = Least Concern DD = Data Deficient NE = Not Evaluated ✪ = Best place to watch

Old World Monkeys

Main image Red-shanked Doucs live in large groups with about twice as many females as males; in captivity they have been seen to care for others' infants.

Introduction: Old World Monkeys

The monkeys of Africa and Asia range in size from tiny talapoins to massive male baboons, living in habitats from arid savanna to tropical swamps and snowy mountains. They differ from New World monkeys in several ways: their tails are not prehensile (able to grasp); they have four fewer teeth than the typical New World monkey; and their nostrils are narrow and point downward—for the latter feature these monkeys and the apes are placed in a branch of the family tree called catarrhines or "narrow noses."

Distinctive teeth

All catarrhine primates—that is Old World monkeys and apes, including humans—possess 32 teeth. This means they have one less premolar in each jaw than New World monkeys. The shape of Old World monkey molar teeth is very distinctive: each molar has four bumps, or cusps, arranged in two ridges called lophs; and each loph has two lobes, or projections. As a result, these teeth are known as bilophodont molars and can easily be distinguished from the simpler molars of hominids, even in the case of a single fossil tooth. The shape of the teeth may be an adaptation to a diet with a high proportion of chewy leaves.

Old World monkeys are today found all over Africa and tropical Asia, reaching as far north as Japan. Until relatively recently they were also found across Europe, but apart from the Barbary Macaques living on Gibraltar at the southern tip of Spain, none live wild there today. We can get some idea of how their extinct European cousins might have behaved by observing the macaques kept in large wooded enclosures at some wildlife parks.

Above A male Dusky or Spectacled Leaf Monkey yawns widely. This behavior may be a result of nervousness or tiredness; it may also serve to remind potential rivals of the owner's sharp teeth.

Evolution and fossil history

The earliest known fossils recognizable as Old World monkeys were found in Africa in 1997 and date back some 15 million years. Discovered on Maboko Island in Lake Victoria, one nearly complete skull was given the name *Victoriapithecus* and is thought to represent an intermediate form that lies in the family tree somewhere between today's apes and monkeys. *Victoriapithecus* is very monkeylike in appearance but has some features that make it resemble a kind of miniature orangutan. Its diet probably consisted mainly of fruit.

Fossil monkeys and prosimians dating from the Miocene Epoch (23–5 million years ago) are rare, and yet there are many fossil apes of numerous species. It seems that the diversity of apes and Old World monkeys then was the reverse of today; over the past 12 million years, ape species were reduced in number to the present-day apes, while Old World monkeys diversified to fill vacant ecological niches. Ironically, the most successful primates of all, humans, are now in the process of reducing the diversity of all other primates by appropriating almost every habitat on Earth and either killing or outcompeting them.

Facing page As night falls, Olive Baboons seek a perch safely out of the reach of predators for the hours of darkness, in Kenya's Masai Mara Nature Reserve.

Above A Blue Monkey stuffs its expandable cheek pouches with food before retreating to a place where it can safely consume its meal later.

Generalists and specialists

All the living Old World monkeys are classed in the family Cercopithecidae, which confusingly means "tailed apes." Soon after this line diverged from the New World monkeys, it divided again into two clear subfamilies—one of specialists and one of generalists. Every Old World monkey species is active during the day and the majority have a larger body size than most New World primates and prosimians.

The generalists in this family are the cheek-pouch monkeys, subfamily Cercopithecinae. Most members of this large and successful group are found in Africa, including the well-known baboons (see pages 112–115) and guenons (see pages 126–131), but it also includes the widely distributed macaques of Asia. The generalists eat almost anything—seeds, fruit, flowers, roots, insects and, in some cases, vertebrates, including birds, reptiles and small- to medium-sized mammals. In addition to cheek pouches, these species have another distinctive anatomical feature—ischial callosities, or seat pads, which are patches of tough, horny skin on the bottom that protects that area when sitting.

The specialists are the leaf-eating monkeys, subfamily Colobinae. There are many species of this group found across both Africa and Asia, but most are of limited distribution and adapted to a specific dietary niche. These monkeys spend most of their time up in the canopy, where they harvest mostly leaves, which they supplement with fruit and flowers.

Cheek-pouch monkeys

Foraging is a dangerous business. Primates plucking leaves or fruit in the forest canopy, or searching for seeds or insects on the ground, must focus their attention on finding food before their neighbor. This leaves them vulnerable to attack. How much better to be able to gather a basket of food, then sit somewhere less exposed to eat? In the absence of the ability to make baskets, Old World monkeys have developed cheek pouches instead.

Watch closely as a troop of macaques gathers seeds or grain or as guenons collect fruit: notice that the monkeys do not pause to chew, but instead rapidly stuff food items into their mouth until their cheeks become more and more distended. They are temporarily storing food in these convenient "shopping bags" and will eat it later at leisure.

Cheek pouches have evolved in several types of mammals, including hamsters and giant rats as well as primates. There are two explanations for this: predator risk avoidance and competition avoidance. Many animals forage in flocks, herds, shoals or—in the case of monkeys—troops. Doing so gives them safety in numbers, because the more eyes there are, the greater the chance of spotting a predator and raising the alarm. If a predator does attack, there are numerous targets, so the chances of being the unlucky one are proportionately reduced. This is why you often see mixed troops of monkeys, just as you see flocks of birds of different species in forests or grassland.

Types of predators

Monkeys are important prey items for four main kinds of predator. Airborne attacks by monkey-eating eagles come, literally, out of the blue. A silent swoop of massive wings and a quick death from specially adapted talons soon follows. Nevertheless, an adult male guenon will sometimes rush at an eagle after a family member has been caught, and sometimes the eagle will be intimidated enough to let go. Stealth attacks by big cats are equally deadly, and Leopards are feared so much that a glimpse of spotted fur is enough to cause panic among a troop of monkeys. But once a Leopard is spotted by a troop of baboons, they will keep barking at it and even mob the cat to chase it away. Large pythons cause panic too if they are spotted coiled around a branch, waiting for an unwary monkey to come within striking distance.

The fourth kind of predator is the most deadly: two of the great apes—Chimpanzees and humans. Chimpanzees hunt monkeys by chasing them through the treetops until one of them makes a mistake and falls or is caught. Some Chimpanzee communities hunt cooperatively, with some members playing the role of drivers who "beat" the prey through the forest while others wait ahead to ambush the prey. Humans, however, traditionally use projectiles such as arrows or darts from blowpipes, in which case the first the troop knows of the attack is when one member keels over and

Below Hanuman langurs, also known as gray langurs, spend four-fifths of their day on the ground, but prefer to sleep in trees, cliffs or buildings.

plummets to the forest floor. Nowadays, most human hunters use shotguns, and the sound of these at least alerts troop members and other animals to the attack.

With all these dangers, it is not surprising that monkeys seek to benefit from the combined vigilance of others. However, the drawback to group foraging is that there are more mouths to feed. Just as you are eyeing up your next mouthful while chewing the last, another member of the troop might step in and snatch it. In a mixed troop, your competitor might be a bigger species. Again, there is a clear advantage to being able to rapidly gather more than you can eat and withdraw to enjoy it later. It is not clear which is the greater benefit—avoiding predators or competition. It is probably a combination of the two, varying according to species and circumstance. Either way, the selective advantage was enough to result in the evolution of cheek pouches in baboons, guenons and mangabeys in Africa and in macaques in North Africa and Asia. Some plants also benefit if their seeds are dispersed via monkeys' cheek pouches. Cheek-pouch monkeys tend to spit out most of the seeds when they eat their fruit, unlike other types of monkey, which swallow most of them. A study of seed dispersal in Japanese Macaques concluded that the rate of germination for larger seeds was better for seeds stored in cheek pouches than for those ingested and passed in the feces.

Macaques

Macaques are among the most successful primates on the planet; there are currently 21 species recognized, and they cover a greater geographical range than any other group of primates except humans. From North Africa to the Himalayas, to Japan and Southeast Asia, their ability to exploit different habitats and live alongside humans has made them more resilient to human-induced change than other, more specialized primates.

The macaque evolutionary line branched off from ancestral baboons about 7 million years ago, and beginning about 5.5 million years ago successive dispersal waves across Asia led to the evolution of three groups of related species within the genus *Macaca*. The largest and most widely dispersed group is the *Silenus-sylvanus* lineage, which consists of 11 species, including India's Lion-tailed Macaque, the Pig-tailed Macaque of Southeast Asia and the Barbary Macaque of North Africa.

Stocky monkeys

Macaques are solidly built, stocky monkeys with bare skin—bright red in some species—on the face and bottom. They are often the easiest and most rewarding monkeys to watch. Because they generally live in large multimale groups, there is usually a lot of activity to observe, such as mothers caring for infants, juveniles chasing and wrestling, and males settling dominance disputes, all spread across an area of several hundred square yards. Macaques are omnivorous, with a liking for fruit, seeds, young leaves, insects, small mammals, lizards and—given half a chance—agricultural crops.

Facing page The Barbary Macaque, Europe's only free-living nonhuman primate, once ranged over much of the southern part of the continent.

Their habit of raiding fields brings macaques into conflict with farmers in some regions, but most have the good fortune to live in proximity with the many human cultures that tolerate or even revere monkeys for religious or other traditional reasons. Hindus worship the monkey deity Hanuman and Buddhists teach reverence for all life, with the result that some macaques even get fed on a regular basis. Macaques are also fed at some tourist viewing sites, to enable visitors to watch their fascinating social behavior.

Europe's only primates

The Barbary Macaque is one of the best-known species because of its colony on the Rock of Gibraltar—the only free-living nonhuman primates in Europe. This colony was established by British sailors, who released pets in 1704 when they took this strategic stronghold. Confusingly, the species is also known as the Barbary Ape or Rock Ape.

The species was once found across much of Europe, but it is now restricted to patches of ancient forest in the Atlas Mountains of Morocco and Algeria, together with the managed colony of around 160 named animals on Gibraltar. In North Africa, these macaques are threatened by habitat loss and poaching for the illegal pet trade, but when left to their own devices Barbary Macaques well adapted to the extremes of this harsh mountain environment. In summer, the temperature in the Atlas Mountains soars to as high as 113°F (45°C), whereas in winter it can plunge to just above 0°F (–18°C).

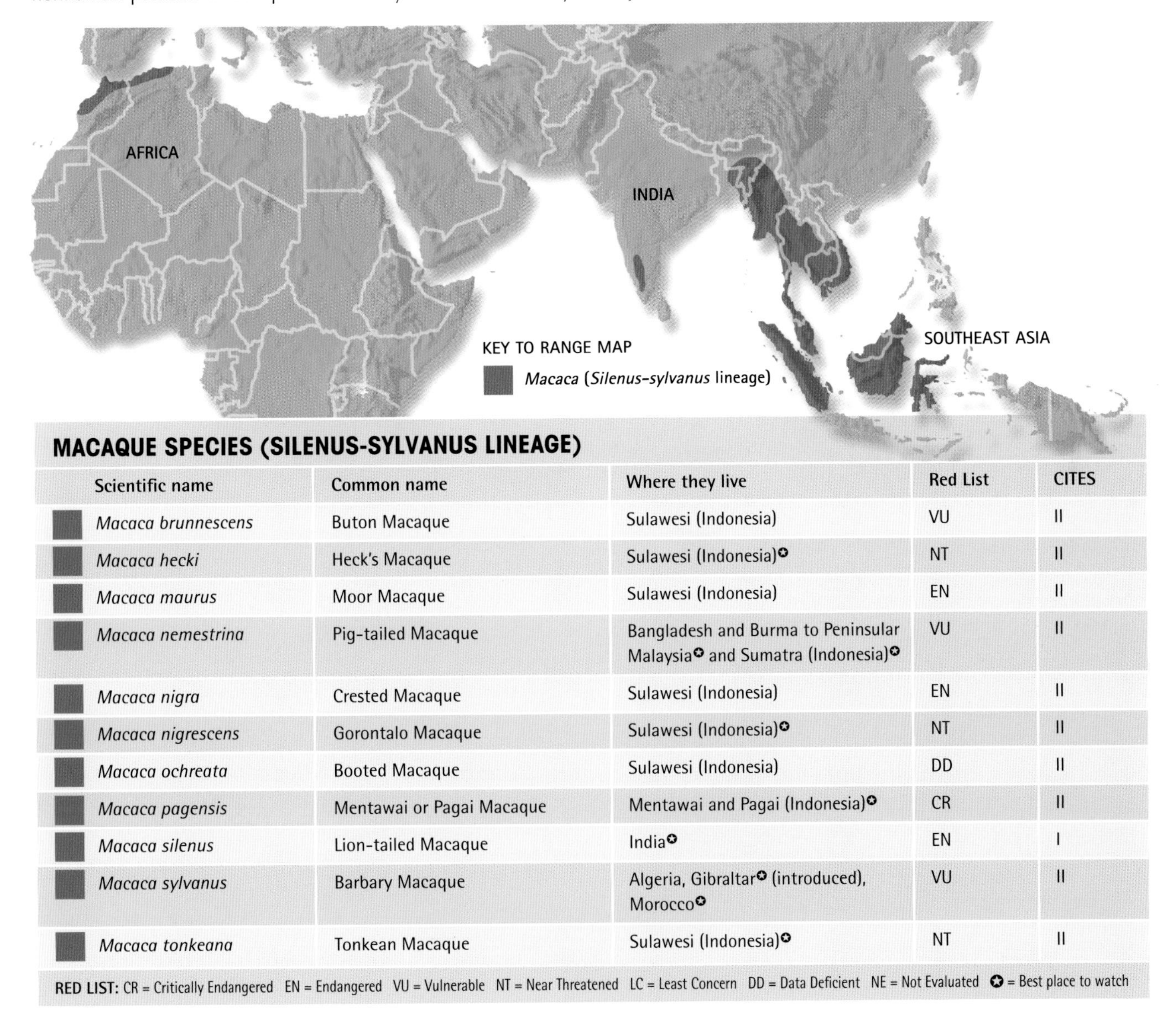

MACAQUE SPECIES (SILENUS-SYLVANUS LINEAGE)

Scientific name	Common name	Where they live	Red List	CITES
Macaca brunnescens	Buton Macaque	Sulawesi (Indonesia)	VU	II
Macaca hecki	Heck's Macaque	Sulawesi (Indonesia)✪	NT	II
Macaca maurus	Moor Macaque	Sulawesi (Indonesia)	EN	II
Macaca nemestrina	Pig-tailed Macaque	Bangladesh and Burma to Peninsular Malaysia✪ and Sumatra (Indonesia)✪	VU	II
Macaca nigra	Crested Macaque	Sulawesi (Indonesia)	EN	II
Macaca nigrescens	Gorontalo Macaque	Sulawesi (Indonesia)✪	NT	II
Macaca ochreata	Booted Macaque	Sulawesi (Indonesia)	DD	II
Macaca pagensis	Mentawai or Pagai Macaque	Mentawai and Pagai (Indonesia)✪	CR	II
Macaca silenus	Lion-tailed Macaque	India✪	EN	I
Macaca sylvanus	Barbary Macaque	Algeria, Gibraltar✪ (introduced), Morocco✪	VU	II
Macaca tonkeana	Tonkean Macaque	Sulawesi (Indonesia)✪	NT	II

RED LIST: CR = Critically Endangered EN = Endangered VU = Vulnerable NT = Near Threatened LC = Least Concern DD = Data Deficient NE = Not Evaluated ✪ = Best place to watch

Left The Tibetan Macaque has mottled skin on the face. Its dense fur insulates against the cold in its mountain habitat.

A spectacular discovery

Given that macaques are such a familiar kind of monkey, scientists were astonished to find an undescribed species in 2004—a century after the last new macaque was described. Named the Arunachal Macaque, it is a large, stocky, brown monkey with a short tail, living only in remote mountain forests of Arunachal Pradesh in northeast India, east of Bhutan. It was, of course, well known to the local Dirang Monpa people, but living at altitudes of 5,000–11,500 ft. (1,600–3,500 m), it had somehow escaped the attention of zoologists.

Often new species are described because one or more subspecies of an already well-known animal are later found to be sufficiently different to warrant being a full species in their own right. The Arunachal Macaque, however, was truly a new discovery to science—the species simply hadn't been seen before by anyone who realized its significance. It was named *Macaca munzala* after the local term *munzala*, which means "deep forest monkey." Fortunately, given its very restricted distribution, the monkey is not hunted for food, but locals report that some are killed in retaliation for crop-raiding. Its habitat is within the western range of the Assamese Macaque, and both these species are in the *Sinica-arctoides* lineage, thought to be the second wave of macaque dispersal across Asia, which resulted in the evolution of six, mostly well-separated species (see table).

The aging process

Macaque species found across the Himalayas tend to be stocky with thick, shaggy fur, which seems appropriate for cold mountains, although the Stump-tailed or Bear Macaque has the same appearance all the way south through Thailand to northern Malaysia. This species has mottled red and black skin on the face, and dark brown fur (apart from infants, which have white fur). As they get older, Stump-tailed Macaques show a similar pattern of baldness to human males, starting with an increasingly high forehead and advancing over the crown, though in the macaques it affects both sexes. The amount of black mottling also increases with age and with exposure to sunlight, so it is easy to recognize older adults. The markings probably also aid recognition between

CHINA

INDIA

SOUTHEAST ASIA

KEY TO RANGE MAP

Macaca (*Sinica–arctoides* lineage)

MACAQUE SPECIES (SINICA-ARCTOIDES LINEAGE)

Scientific name	Common name	Where they live	Red List	CITES
Macaca arctoides	Stump-tailed Macaque	Southeast Asia, China, Thailand✪	VU	II
Macaca assamensis	Assamese Macaque	Nepal✪, India✪ to Vietnam and China	VU	II
Macaca munzala	Arunachal Macaque	Arunachal Pradesh (northern India)	NE	II
Macaca radiata	Bonnet Macaque	India✪	LC	II
Macaca sinica	Toque Macaque	Sri Lanka✪	VU	II
Macaca thibetana	Tibetan Macaque	China	NT	II

RED LIST: CR = Critically Endangered EN = Endangered VU = Vulnerable NT = Near Threatened LC = Least Concern DD = Data Deficient NE = Not Evaluated ✪ = Best place to watch

troop members. Stump-tailed Macaques live in groups of up to 40 individuals, in which female hierarchies are very strong. The eponymous tail is an important signaling device within Stump-tail society. Signaling your intent can avoid a fight, so "tail down" means submission or fear, and may be accompanied by a teeth-chattering face to show appeasement or willingness to groom; an upright tail is a sign of confidence, while one that is curled up indicates excitement.

Tourists visiting the temples and historic sites of Sri Lanka will be familiar with the more slender, southernmost species in this lineage—the Toque Macaque. Thanks to its bright red face and the curious pattern of hair on top of its head, radiating out from the center of the crown, this primate is unmistakable. Like the closely related Bonnet Macaque of southern India, Toque Macaques swim well.

In both Toque and Stump-tailed macaques, fruit and seeds make up the bulk of the diet, and they are supplemented by other plant parts and animal prey, including insects, lizards and frogs. Toque Macaques also catch and eat small birds and mammals. Like other macaques, they raid dustbins, crops and food stores, but because they live on a Buddhist island, they are persecuted much less than most macaques elsewhere.

PRIMATE PROFILE
MACAQUES

SIZE Head and body length: 14½–30 in. (37–76 cm); weight: 5½–39 lb. (2.5–17.5 kg)

APPEARANCE Short coat (color varies according to species); broad face; short body; limbs of equal length

HABITAT Wide range of upland and lowland forest types, including evergreen, broadleaf, bamboo and mangrove

DIET Fruit, seeds, leaves, flowers, buds, insects, small animals

LIFE HISTORY Gestation: about 6 months; sexual maturity: male not known, female 3–7 years; life span: up to 33 years

BEHAVIOR Active during the day; tree-living; variable social structures, often with strict hierarchies; communicates through calls, facial expressions and gestures; grooms

Below Female Toque Macaques, with red faces and a circular patch of hair atop, groom one another and their infants during a rest period in Sri Lanka.

Above These Japanese Macaques save energy by resting and socializing in geothermal hot springs during Japan's snowy winters. This behavior is seen in only a few troops that have access to such springs.

It is an iconic image: a red-faced monkey up to its neck in a hot spring, with snowflakes settling on its thick head fur. This is the popular view of the Japanese Macaque, or snow monkey. What is less well known is that only one or two troops of macaques have taken to bathing in hot water, in Shiga Heights, central Japan; everywhere else, they just huddle together on the ground in cold weather—the lower the temperature, the larger the huddle.

Macaque handouts

High on a hillside above the old city of Kyoto, a man with a bucket is scattering grain. Dozens of shaggy monkeys are rapidly transferring this bounty, one grain at a time, into their cheek pouches, using a quick picking action that alternates between the left and right hand. Tourists snap photos and scientists gather data, but the macaques seem oblivious. Scenes like this have been the norm in the Arashiyama district for more than 50 years, and as a result, the local macaque population is among the best known of any species. The monkeys are entirely free-living and spend a lot of their time in the surrounding forest. When food in the forest is plentiful, only a few will turn up for a free meal. But when forest food is scarce, numbers at the provisioning station rise, making this one of the easiest places on the planet to watch friendly, habituated nonhuman primates—all within walking distance of a major metropolis.

Rainforest and swamp macaques

At the other extreme, Long-tailed Macaques are found in tropical rainforests and swamps from Burma (Myanmar) south to Indonesia and the Philippines. Their eponymous long tails exceed the length of head and body combined; they have a prominent facial fringe of hair and a distinctive patch of white in the corner of each eye. These macaques are known for their swimming abilities and, as their alternatve name (Crab-eating Macaque) suggests, they forage in mangroves and swamps for crabs. They also catch other invertebrates and frogs.

female's mother are critical, and arguments are settled wit aggression and bared teeth, followed by reconciliation throug grooming and embracing. The females may not be dominant ove the males, but they form stable hierarchies in which daughter inherit their mother's rank; the males change groups every fev years and live on the edge of the large troops. Young males avoi conflict with adult males in the group by leaving at about fiv years old, because sexual maturity makes them more of a threat Some of these males remain alone for as long as a year or two, bu others join up in bachelor groups. Eventually, they join a breedin group as subordinate males and, as they gain age and experience they begin to aggressively challenge the alpha male—thereb gaining the breeding rights to the group's females. The alpha mal does not sire all the offspring, however. In every species studied females also mate with male visitors, and these extra-grou liaisons result in a significant number of offspring.

bove A female Rhesus Macaque grooms an infant. After the first few weeks, naternal duties may be shared by female relatives, especially by those who need o learn. Researchers have found that female infants receive more care than males.

apanese and Long-tailed Macaques are two of the four species hought to have resulted from the third wave of macaque lispersal across Asia; the other two species are the Rhesus and aiwanese (formerly Formosan) macaques. Collectively, hese four species have the greatest distribution of ny nonhuman primates. They are among the most umerous primate species too, apart from the aiwanese Macaque, which is restricted to the island f Taiwan and considered threatened.

Political alliances

Research into Rhesus and Japanese macaque societies indicates hat their troops are a hotbed of political alliances and competitive social relations between females. Hierarchy and the status of each

KEY TO RANGE MAP

Macaca (*Fascicularis* lineage)

MACAQUE SPECIES (FASCICULARIS LINEAGE)

Scientific name	Common name	Where they live	Red List	CITES
Macaca cyclopis	Taiwanese Macaque	Taiwan✪	VU	II
Macaca fascicularis	Long-tailed or Crab-eating Macaque	Bangladesh, Burma, Nicobar Islands, Philippines✪, Malaysia✪, Indonesia✪	NT	II
Macaca fuscata	Japanese Macaque	Japan✪	DD	II
Macaca mulatta	Rhesus Macaque	Afghanistan and India✪ to Thailand✪ and southern China	NT	II

RED LIST: CR = Critically Endangered EN = Endangered VU = Vulnerable NT = Near Threatened LC = Least Concern DD = Data Deficient NE = Not Evaluated ✪ = Best place to watch

Baboons

Revered by the ancient Egyptians, photographed by tourists, studied by scientists and hated by farmers, baboons have always had mixed fortunes in their relationships with humans. Nevertheless, they remain a big African success story. Their life on the savanna has been examined closely in attempts to draw parallels with another primate that moved from the forest out onto the plains—*Homo sapiens*.

Above The threat display of the male Hamadryas Baboon, whose long, bushy mane makes him look larger than he is, is used to impress both rivals and females.

Baboons are also known as "dog-faced monkeys" because of their elongated muzzle, and they are intelligent, sociable and fascinating to watch. They are unusual monkeys, not least because they spend most of their time on the ground. They have a savvy, streetwise demeanor that suggests they know how to handle themselves. Adult males have the weapons (long, sharp canine teeth) and the muscles (they are twice the size of females), and they don't so much walk as swagger. Even when backing down from a more dominant troop member or a larger species, they look as though they are already planning the next move.

This behavior has been observed for thousands of years; in ancient Egypt, Hamadryas Baboons were thought to represent the god of wisdom, Thoth, also known as Tahuti, and four sacred baboons were often depicted adoring the Sun in his temple at Hemmopolis. Anyone who has watched a troop of baboons sitting with arms out, calling to one another, soaking up the early morning rays before starting the day's foraging, will understand how this notion came about.

Power struggles

When field studies of baboons began in Kenya in the early 1970s, it was thought that baboon troops were run by aggressive males that fought their way to the top. The results of the research, however, showed that baboon society is far more complex than that, and males are more likely to use subtle or not so subtle political tactics rather than brute force to improve their social standing. They win friends by protecting and grooming individuals, who they can then call on to help in a dispute. The males who are best at making friends and getting along with youngsters turn out to get more mating opportunities than aggressive males who seek to dominate everyone. To avoid a fight, a socially adept male might hold onto an infant or a female, knowing the attacker will be inhibited from harming the weaker members of the troop.

Hamadryas Baboons, found in the Horn of Africa and across the Red Sea in Yemen and Saudi Arabia, show four levels of social organization. A family consists of one male and a number of females; two to four families with related bachelors make a clan; two or three clans form a band of about 60 individuals; and several of these bands join up in a troop, or even a multi-troop aggregation, to sleep on cliffs or forage together.

Social structure

Female Olive and Yellow baboons form matrilines (a system in which an individual's place in society is based on the lineage of the mother). Daughters remain within their natal troop and tend to inherit the status of their mother. Males transfer out of the troop they were born into at about four years of age. In Hamadryas Baboons, the situation is reversed: males stay within their natal clan while females transfer out at about three-and-a-half years. Female baboons first become sexually active at about four or five years of age and begin to show the pink sexual swellings that advertise their receptivity to potential mates. Once they become pregnant, the thickened skin that forms the seat pads on the buttocks becomes red in the females of several baboon species.

When on the move, a baboon troop adopts a well-organized formation, with the dominant male in front, followed by the dominant female. Younger males form a flank on either side and bring up the rear, protecting the females and juveniles in the center. As well as keeping a lookout for danger from outside the troop, low-ranking animals are constantly alert to the whereabouts of dominant individuals that might displace them from where they are eating, resting or even mating.

Below Young primates, such as this young Olive Baboon, live in complex societies. Their success depends on learning appropriate behavior.

Taxonomic riddles

Baboons are present throughout sub-Saharan Africa, except in the densest rainforests and driest deserts. As in so many primate groups, the number of species in a genus depends on whether you follow the taxonomic "splitters" or "lumpers." Some authors in the latter category lump different forms together and accept only two full species, the Gelada and Hamadryas baboons, of which the latter, *Papio hamadryas*, is given many subspecies. Other authorities prefer to split different forms, and consider that at least five of those subspecies warrant full species status (see the table on page 115). There has also been a suggestion that the various kinds of *Papio* might be divided into desert types and savanna types, but this does not stand up to close scrutiny. The situation is further confused by the tendency for some of the five species to interbreed and form hybrids along the boundaries of their ranges.

The genus *Theropithecus* contains only a single species, the Gelada Baboon of the Ethiopian Highlands. It is the sole survivor of a number of larger, more widespread species that were hunted by early Hominids. The Gelada differs from other baboons by having smaller incisors and larger cheek teeth. This baboon also has very short index fingers and a more opposable thumb than other Old World primates. The most obvious difference in the living animal, however, is the cape of long hair over the shoulders and the bright red skin of the cleavage and up under the chin, forming a red hourglass shape. This is most dramatic in the males; the brightness of the red skin denotes status and is enhanced by surrounding white fur. It becomes more pronounced in the females when they come into estrus, and numerous swollen red bumps, or vesicles, protrude from the skin on their chest. This is thought to be a better advertisement for ovulation than a red rump in a species that spends so much time sitting down.

Large gatherings

Geladas also have a multitiered society, with troops of up to 300 consisting of bachelor groups and clusters of related females all serviced by one dominant male. He remains the boss for as long as he is able to fight off challenges from any ambitious younger males.

PRIMATE PROFILE
BABOONS

SIZE Head and body length: 21½–34 in. (54–86 cm); weight: 33–62 lb. (15–28 kg); males usually larger than females

APPEARANCE Coarse fur (color varies according to species); long face; long snout; short tail; stocky body

HABITAT Gallery forest; rainforest; montane grassland; wooded savanna; scrub

DIET Fruit, seeds, leaves, flowers, roots, grass, small animals

LIFE HISTORY Gestation: about 6 months; sexual maturity: male 5–7 years, female 4.5 years; life span: up to 45 years

BEHAVIOR Active during the day; spends most time on the ground, but sleeps in trees or on cliffs; lives in complex groups of variable social structure; grooms

Once he retires, the alpha male stays in the troop as a peripheral, nonbreeding male. Sometimes a number of troops gather at a feeding site in a temporary congregation of hundreds of animals.

Geladas have a specialized diet and spend half their waking hours sitting on their buttock pads plucking grass stems and clovers in the open mountain pastures. In the dry season they eat roots, bulbs and rhizomes. As there is no shortage of grass, there is little point in troops acting defensively, nor is there much sign of cohesion within the troop. There is, however, a lengthy period of socializing each morning, when relationships are cemented by social grooming. At dusk, the baboons retire to sleep on spectacular cliffs in gorges, where they are safe from predators.

Staying in touch

Communication in large gatherings is important, and half of Gelada Baboon vocalizations are contact calls. When two Geladas approach one another, the dominant one may signal a threat by flashing the eyebrows, exposing the pale skin of the eyelid. If the submissive one becomes frightened, it will express its fear with a "lip flip" grimace, which entails flipping up the upper lip so it covers the nostrils, exposing the red gum and teeth.

Sadly, Geladas are killed for their crop raiding and have lost much of their former range to agriculture. This has pushed them into areas previously occupied only by Olive Baboons, and despite belonging to different genera, these two species have interbred.

Below Troops of Gelada Baboons are led by a large alpha male; breeding groups may band together to form gatherings of up to 300.

BABOON SPECIES

Scientific name	Common name	Where they live	Red List	CITES
Papio anubis	Olive Baboon	Mali✪ to Rwanda✪, Uganda✪, Kenya✪	LC	II
Papio cynocephalus	Yellow Baboon	East Africa✪ to Zambia✪ and Angola	LC	II
Papio hamadryas	Hamadryas Baboon	Ethiopia✪, Saudi Arabia, Somalia, Yemen✪	NT	II
Papio papio	Guinea Baboon	Gambia✪, Guinea✪, Guinea-Bissau, Mauritania, Senegal✪, Sierra Leone	NT	II
Papio ursinus	Chacma Baboon	Namibia✪ to Zimbabwe✪	LC	II
Theropithecus gelada	Gelada Baboon	Ethiopia✪	LC	II

RED LIST: CR = Critically Endangered EN = Endangered VU = Vulnerable NT = Near Threatened LC = Least Concern DD = Data Deficient NE = Not Evaluated ✪ = Best place to watch

Mandrill and Drill

Mandrills are famous for being the most colorful of all mammals, and they are also the world's largest monkeys. They and the slightly smaller Drills are certainly spectacular—males of both species display an eyecatching array of sexual adornments. Until recently, however, the social life of these distinctive primates was largely hidden by their dense forest habitat, which makes it difficult to study them in the wild.

In 1871, Charles Darwin wrote that: "No other member in the whole class of mammals is colored in so extraordinary a manner as the male Mandrill." He had a point. A male Mandrill is instantly recognizable, with his brilliant red nose providing a stark contrast to the electric-blue grooved flanges in which it sits. Huge canines project downward either side of his golden beard. The other end of the Mandrill's body also reveals an extravagant display of color: his hairless rump ranges from blue to purple, and his genitals are bright blue and red.

An adult male Drill is perhaps even more impressive from the rear. His buttocks are adorned with a wash of lilac, purple, red, white and yellow. His face, although it lacks the color of the Mandrill's, is equally striking, if not a little menacing. Bright orange eyes stare out from a black, shiny, contoured face, framed with white fur.

PRIMATE PROFILE
MANDRILL AND DRILL

SIZE Head and body length: 22–32 in. (56–81 cm); 25–66 lb. (11.5–30 kg); males larger than females

APPEARANCE Males: brightly colored snout and/or rump (according to species); females: dull-colored snout and beard; long snout

HABITAT Primary and secondary forest; gallery and lowland rainforest; montane and coastal forest

DIET Fruit, seeds, leaves, bark, stems, fungi, palm nuts, roots, small animals

LIFE HISTORY Gestation: about 7 months; sexual maturity: male not known, female 3.5 years; life span: up to 46 years

BEHAVIOR Active during the day; tree- and ground-living; lives in large groups with strict hierarchies

Facing page The startling colors of a male Mandrill are affected by his social standing—when he becomes dominant, hormonal changes make the colors brighter. Once he is deposed, he literally fades into the background again.

Sexual selection

The remarkable features of the Mandrill are the result of generations of females choosing the most brightly colored males as mates. Males compete fiercely for the right to reproduce. During the breeding season males fight, scent-mark, threaten each other and grunt to demonstrate their strength and confidence. Bigger males—and bigger teeth—make for greater success, leading to the sexual differences in form we see today: an adult male may be twice as big as a female, and his canine teeth may reach 2½ in. (6.5 cm) in length. Males build up fat reserves to help them through these times, when foraging becomes a lower priority than fighting for status.

Male Drills and Mandrills leave their family group before reaching sexual maturity, while females usually stay in the group in which they were born. Due to the difficulties involved in studying them, their seemingly complex society is not fully understood.

Below Sexual selection often leads to the evolution of similarities between the appearance of the front end and the back end among primates, but seldom in the colors displayed by the male Mandrill.

Above The male Drill may lack the arresting colors of his cousin the Mandrill, but the sculpted, masklike face and intelligent gaze give him a commanding presence. Drills have prominent bony ridges on the sides of their nose; they are not grooved, unlike the facial ridges of the Mandrill.

Various group structures have been reported and interpreted in different ways. Groups containing one, several and no males have all been observed, as have solitary males, and group sizes range from 10 individuals up to amazing 1,000-strong troops known as hordes. Whatever the group's social structure, it seems that such large gatherings of Drills or Mandrills are created by the congregation of a number of smaller troops.

Both Drills and Mandrills maintain group cohesion in dense forest by means of frequent deep grunting and high-pitched "crow calls." Scent-marking also sends signals to other members of the species—especially during the breeding season, when males secrete a heavily scented liquid from glands on their chest, rubbing their chests against trees to leave the scent.

Down to earth

The Mandrill and Drill are "bottom feeders" of the forest. Both species spend most of their time on the ground scouring the forest floor for fallen foods. They rummage through the leaf litter to find hard-shelled fruits and seeds, which they prise open using their dextrous hands and sharp teeth. They roll logs over and rip the bark away to reveal nutritious fungi and invertebrates. No stone is left unturned. When the monkeys find a pond, they stand at the

Below right This young male Drill will have to decide whether to emigrate, to live alone or to make a bid for leadership of his group, in which he ranks low.

edge, pulling out dead leaves and twigs, and picking out fallen fruits and seeds. Young Drills often take this time as an opportunity to play, frolicking and splashing in the water. All of these monkeys climb into the trees to sleep.

Under threat

The impenetrable habitat, wide-ranging lifestyle and shyness of the Mandrill and Drill make them difficult primates for researchers to study, but unfortunately their noisy grunting and large troops make them easy targets for hunters to locate. Hunting these large-bodied primates for bushmeat has become a profitable business, and bushmeat hunters may wipe out entire populations.

Commercial logging and forest clearance for agriculture continue to destroy the habitat of the Mandrill and Drill; the Drill is now one of the most endangered African primates and the Mandrill is listed as vulnerable. Efforts are underway to halt the decline. Cameroon has begun enforcing wildlife law, and in 2002, Gabon set aside 10 percent of the country as national parks. Research may shed more light on these colorful primates and help to ensure they survive.

Above This family group of Drills lives in a forested sanctuary in southeast Nigeria, in which victims of the bushmeat trade are rehabilitated before being returned to the wild into protected forests.

MANDRILL AND DRILL SPECIES

Scientific name	Common name	Where they live	Red List	CITES
Mandrillus leucophaeus	Drill	Cameroon✪, Nigeria✪	EN	I
Mandrillus sphinx	Mandrill	Cameroon to Gabon✪ and Congo	VU	I

RED LIST: CR = Critically Endangered EN = Endangered VU = Vulnerable NT = Near Threatened LC = Least Concern DD = Data Deficient NE = Not Evaluated ✪ = Best place to watch

Mangabeys

Mangabeys are large, elegant monkeys that were given their name by accident, because the first shipment of these monkeys to Europe was mislabeled as originating from Mangabe in Madagascar. In reality they are found in the tropical evergreen and swamp forests of West and Central Africa, extending east as far as the Tana River in Kenya.

Once classified as a single genus, the mangabeys are now split into three genera: the crested mangabeys (*Lophocebus*), the white-eyelid mangabeys (*Cercocebus*) and the recently discovered Highland Mangabey (*Rungwecebus*). Discovered by two research teams working independently, the Highland Mangabey, or Kipunji as it is known locally, is found only in the highland forests of Tanzania and was first described in 2005.

Genetic analysis concluded that the Highland Mangabey is closely related to baboons, but differences in its skeleton and anatomy made it distinct from them. Its distinctive "honk bark" call distinguishes it from other mangabeys, which use "whoop gobbles" to communicate. The "whoop" gets the attention of other mangabeys; the "gobble" tells them the identity and location of the caller.

The two crested species of mangabeys have some similarities with baboons, while the six white-eyelid species have more in common with Drills and Mandrills. As a result, the crested monkeys are often referred to as "baboon mangabeys," and the white-eyelid group as "drill mangabeys." Crested mangabeys spend nearly all of their time in the canopy of the evergreen forests in which they live. They generally feed on fruit, with a particular liking for figs, but will just as happily devour almost anything else that they find as they clamber through the trees. This may include insects and seeds, particularly those with hard shells.

In contrast to their crested cousins, white-eyelid mangabeys prefer to spend their time on the forest floor. Here they scour the undergrowth for tasty morsels, in much the same way as the closely related Mandrills. They often associate with canopy-feeding

PRIMATE PROFILE
MANGABEYS

SIZE Head and body length: 15–35 in. (38–89 cm); weight: 10–46 lb. (4.5–21 kg)

APPEARANCE Coat color varies according to species; pale eyelids; large cheek pouches; long, slender bodies

HABITAT Wide variety of forest types: tropical rainforest, dry forest, swamp and mangrove forest, semideciduous forest

DIET Fruit, seeds, leaves, flowers, bark, small animals

LIFE HISTORY Gestation: about 6 months; sexual maturity: male 5–7 years, female 3 years; life span: up to 33 years

BEHAVIOR Active during the day; tree- and ground-living; forms multimale and multifemale groups; associates with other monkey species

guenons. Any fallen fruit, seeds, fungi or invertebrates are eaten. Rotting tree trunks and branches provide plenty of nutritious snacks, and are ripped apart in search of such treats. Studies of mangabeys have shown that they can figure out the places to search for fruit by using the weather as a guide, leading to speculation that this might be how humans developed advanced cognitive skills.

Different social systems

Male white-eyelid mangabeys are much larger than the females. This is due to sexual selection, in which generations of males competed for the chance to mate and the most successful—usually the largest—succeeded. When the female is ready to mate, she advertises the fact with swelling around her sexual organs. Males live either in a group with females and their offspring, or roam between groups, associating with females only to mate.

Crested mangabeys breed throughout the year. The single infant is born after six months and is carried by the mother until it can clamber through the branches by itself. Female mangabeys of all species usually stay with their families throughout their life, but males may switch troops once or more.

Threatened populations

Like other large mammals, mangabeys are among the first to disappear when forests are disturbed by encroaching human settlements. Commercial logging on a large scale continues to destroy and fragment their habitat, while hunting for bushmeat also depletes populations. In addition, crop raiding by mangabeys in search of an easy meal often brings this group of primates into conflict with farmers, so local people may kill them to protect their livelihoods.

MAIN MANGABEY SPECIES

Scientific name	Common name	Where they live	Red List	CITES
Cercocebus agilis	Agile Mangabey	Cameroon✪, Congo✪, DRC, Eq. Guinea, Gabon✪	NE	II
Cercocebus galeritus	Tana River Mangabey	Kenya✪	CR	I
Cercocebus torquatus	White-collared Mangabey	Nigeria✪ to Angola	NT	II
Lophocebus albigena	Gray-cheeked Mangabey	West to central Africa, Uganda✪	LC	II
Lophocebus aterrimus	Black Mangabey	Central Africa	NT	II
Rungwecebus kipunji	Highland Mangabey or Kipunji	Tanzania✪	NE	II

RED LIST: CR = Critically Endangered EN = Endangered VU = Vulnerable NT = Near Threatened LC = Least Concern DD = Data Deficient NE = Not Evaluated ✪ = Best place to watch

Patas Monkey and Swamp Monkey

These two very different but closely related monkeys illustrate how adaptable the basic primate form can be under different evolutionary pressures. Patas Monkeys are slender, long-limbed, fast-running primates that lead a terrestrial life in open savanna woodland and grassland. Swamp Monkeys at first glance look like forest guenons, but have webbing between their fingers and toes to help them move through the murky waters of the Congo Basin.

Above In the wild, Allen's Swamp Monkey (pictured here in captivity) licks the nectar, and thereby pollinates, *Daniella pynaertii*, a flowering tree.

Swamp Monkeys

Very little is known of the elusive Swamp Monkey, full name Allen's Swamp Monkey. When it was first discovered in 1907, the species was placed in the same genus as guenons (*Cercopithecus*). In the 1970s, however, it was reassessed and given its own genus, *Allenopithecus*. Swamp Monkeys are found exclusively in Central Africa, on either side of the River Congo. They are semiterrestrial, and live only in swamp or gallery forests, foraging on the ground and in shallow waters during the day, and climbing trees at night to sleep. Swamp Monkeys are one of the few primates known to swim regularly. This ability is used as a means of escape from a pursuing predator, which, one might think, must be taken aback when its primate prey makes a sudden leap into the water.

The limited observations that have been made of Swamp Monkeys suggest that they live in groups of around 40 individuals, made up of both males and females. Group members keep tabs on each other's positions in the forest by grunting. Groups may also spend time with other primates, such as Red-tailed Guenons and Mona Monkeys (see Guenons, pages 126–131), and with more eyes to spot predators, such partnerships may benefit all species concerned. Most of the Swamp Monkey's diet is made up of fruit, supplemented by pith, roots and animal prey such as beetles and worms. During the dry season they have been observed scooping fish hatchlings from muddy pools in riverbeds.

Single infants are thought to be born around June, but little else of the Swamp Monkey's reproduction and social system is known. They are less endangered than many other primates, but their restricted range and dependence on the River Congo system makes them vulnerable to habitat loss and hunting from boats.

Patas Monkeys

With their slim bodies and long, graceful limbs, Patas Monkeys can run at extraordinary speeds of up to 34 mph (55 km/h), making them the fastest of all primates. They are widely distributed in savanna woodland, from West to East Africa on the northern fringe of the tropical rainforest belt. Some authors believe that up to four subspecies of Patas Monkeys may exist, although others think these are just geographical variations.

Patas Monkey groups contain up to 70 members, often consisting of one lucky male, many females and their offspring. When a young male reaches puberty, he will leave his group, sometimes taking a brother along too. They may then join together with other roving males until they are able to overthrow the resident male of another group. They may attempt this during the breeding season, when males often join other groups and try to sneak matings with females. At such times, the group may contain other males for a while, the alpha male being too preoccupied with mating to chase them away. Bonds between females are strong, and they spend a lot of time grooming each other. When a baby is born, other females in the group help to care for it—especially if it is a female. The resident male spends his time on the lookout for predators.

PRIMATE PROFILE
PATAS MONKEY

SIZE Head and body length: 19½–35 in. (49–88 cm); weight: 8¾–29 lb. (4–13 kg); males much larger than females

APPEARANCE Reddish brown back and sides, gray-white underparts; distinctive mustache; long, slender limbs

HABITAT Savanna; acacia woodland

DIET Acacia trees, fruit, seeds, grass, birds' eggs, insects, lizards

LIFE HISTORY Gestation: 5–6 months; sexual maturity: male 4 years, female 3 years; life span: up to 22 years

BEHAVIOR Active during the day; mainly ground-living; forms multifemale/one-male groups; females are dominant and territorial

To avoid revealing their location to predators, Patas Monkeys do not make much noise. Instead, they keep visual contact with each other, frequently standing on two legs and scanning the area to look for danger and check the wherabouts of other group members and rival groups. Although they do not defend ranges, females are aggressive toward other groups if encountered. At sunset, each Patas Monkey climbs into its own tree to sleep.

Below An infant Patas Monkey clings to an adult female's belly. Female infants receive much more attention from other females than male infants do. Young males must make do with playing with each other on the group's periphery.

Wide-ranging primates

Of all primates, Patas Monkeys have the largest home ranges in relation to the total mass of the group in the territory, which may cover up to 15 square miles (40 sq km). The monkeys travel up to 2½ miles (4 km) a day in search of widely dispersed food. Their diet varies according to what is available through the African seasons but may include seeds, a little fruit and large amounts of gum and insects. The monkeys catch grasshoppers in long grass and bite the swollen thorns of prickly acacia trees to get at ants inside the galls. They have even been seen catching fish from small ponds.

Crops provide easy meals and this has made Patas Monkeys unpopular with farmers, who often kill them. They are also hunted for meat, for use in traditional medicine and to be sold in the pet trade. Although they seem able to cope with the current level of hunting pressure, there are fears that as human populations continue to expand, so will the pressure on this beautiful species.

KEY TO RANGE MAP

Allenopithecus nigroviridis

Erythrocebus patas

AFRICA

PATAS MONKEY AND SWAMP MONKEY SPECIES

Scientific name	Common name	Where they live	Red List	CITES
Allenopithecus nigroviridis	Allen's Swamp Monkey	Congo, DRC	NT	II
Erythrocebus patas	Patas Monkey	West to East Africa (Ghana✪, Kenya✪)	LC	II

RED LIST: CR = Critically Endangered EN = Endangered VU = Vulnerable NT = Near Threatened LC = Least Concern DD = Data Deficient NE = Not Evaluated ✪ = Best place to watch

Above Angolan Talapoin Monkeys are good swimmers and seldom venture more than 500 yards (450 m) from a river.

Vervet and Talapoin Monkeys

The Vervet is Africa's most widely distributed primate species, although some authors think there are four to six closely related species. Male Vervets have a distinctive blue scrotum, with a red penis surrounded by white fur. The talapoins are the smallest Old World monkeys. They look very like the squirrel monkeys of South America, except for the male's blue scrotum.

Any visitor to the African savanna will be familiar with Vervet Monkeys, known in different areas as Grivet, Tantalus or Green Monkeys. Their attractive black face surrounded by a fringe of pale fur makes them unmistakable, and their cheeky opportunism in stealing food or using vehicles as a perch makes them popular with tourists but a nuisance to hotel managers and local farmers. In fact, Vervets were until recently classed as vermin in South Africa and still are in Uganda, along wih baboons. They are common in savanna woodland and forests throughout sub-Saharan Africa, but are absent from deserts and the dense rainforest of the Congo Basin.

The success of these monkeys is down to their supreme adaptability: they eat a wide range of foods depending on what is available, including fruits, seeds, leaves, tree gum, insects and other small animals. One population in a mangrove swamp eats mainly fiddler crabs, and a montane form in Ethiopia eats bamboo.

The wide distribution of the Vervet Monkey creates a conundrum for primate taxonomists, because there is gene flow between most neighboring populations, and yet there are obvious differences between distant populations. Until *Chlorocebus* was proposed as their genus in 1989, Vervets were in the same genus as guenons, *Cercopithecus*. Not everyone has adopted the new genus, however, and the number of species and subspecies is still disputed (22 subspecies have been described). The Vervets themselves, however, have a clear understanding of who is who in their world.

The size of Vervet Monkey groups varies from 5 to 76, with 50 percent more females than males. Playback experiments of vocalizations reveal a surprisingly sophisticated society; Vervets can not only identify which individual is calling, but also whether he or she is from their own or a neighboring group, and even what his or her position in the group's dominance hierarchy is.

Early warning system

Vervets have distinct alarm calls for different predators, and each call elicits a different reaction. If a monkey shouts the call "SNAKE!," all the monkeys stand on their legs and look around on the ground; if the call "LEOPARD!" is given, they all rush to climb a tree to the thinnest branches, beyond the big cat's reach; if the call is "EAGLE!," the monkeys immediately scan the sky.

Talapoins have a much more restricted distribution than Vervet Monkeys, in primary, secondary and swamp forests, to the north and south of the mouth of the Congo River. Their two populations were recently recognized as separate species, the Gabon Talapoin having shorter, more finely banded yellow and black fur and a shorter tail than the Angolan species. Sometimes known as Dwarf Guenons, they have catholic tastes similar to those of Vervets,

PRIMATE PROFILE
TALAPOIN MONKEYS

SIZE Head and body: 13–18 in. (32–45 cm); tail: 20 in. (50 cm); weight: 1⅗–2¾ lb. (0.75–1.3 kg); males larger

APPEARANCE Gray-green coat on top, whitish belly, short-snouted, round head, hairless face

HABITAT Primary, secondary, gallery, mangrove and lowland swamp forest, usually close to a river

DIET Fruit, leaves, flowers and animal prey, including insects, reptiles and freshwater shrimps

LIFE HISTORY Gestation: average 5 months; sexual maturity: male 9½ years, female 4 years; life span: up to 28 years

BEHAVIOR Active during the day; tree-living; good swimmers; live in large groups

Below Widespread in savanna and woodland habitats across sub-Saharan Africa, Vervet Monkeys are a familiar sight for safari tourists.

VERVET AND TALAPOIN MONKEY SPECIES

Scientific name	Common name	Where they live	Red List	CITES
Chlorocebus aethiops	Vervet, Green, Tantalus or Grivet Monkey	Gambia✪ to Ghana✪, East Africa✪ (Kenya to South Africa)	LC	II
Miopithecus ogouensis	Gabon Talapoin	Cameroon to Congo, Gabon✪	NE	II
Miopithecus talapoin	Angolan Talapoin	Angola, Congo	LC	II

RED LIST: CR = Critically Endangered EN = Endangered VU = Vulnerable NT = Near Threatened LC = Least Concern DD = Data Deficient NE = Not Evaluated ✪ = Best place to watch

AFRICA

KEY TO RANGE MAP
- *Chlorocebus aethiops*
- *Miopithecus ogouensis*
- *Miopithecus talapoin*

and also have a reputation for crop raiding and stealing manioc roots left out to soak by villagers. This ready food supply affects their social structure: talapoin group size is normally 40 to 50 individuals, but in areas near human habitation it rises to twice that, and three times the density of other groups.

When feeding in dense foliage, talapoins keep in touch with frequent contact calls. Females say *"coo"* before and after every move, and at times of excitement (for example, after an alarm or when entering a laden fruit tree) the whole troop gives a chorus of every kind of call in its repertoire.

PRIMATE PROFILE
VERVET MONKEY

SIZE Head and body: 12–24 in. (30–60 cm); weight: 3¼–14 lb. (1.5–6.5 kg); males larger than females

APPEARANCE Short, pale brown to olive fur; pale underparts; black face fringed by pale fur

HABITAT Wooded savanna and marginal habitats near water, including lowland swamps, semi-arid terrain; montane forest

DIET Fruit, seeds, leaves and animal prey, including invertebrates, reptiles, birds and mammals

LIFE HISTORY Gestation period: average 5 months; sexual maturity: male 5 years, female 4 years; life span: up to 31 years

BEHAVIOR Active during the day; ground- and tree-living; lives in groups

Guenons

Crashing branches and high-pitched chirrups are often the first indication of a troop of guenons—beautiful, colorful forest monkeys, found across sub-Saharan Africa. They exhibit a wide range of fur colors and facial patterns, and there are between 23 and 36 species, depending on which classification is used. This diversity is linked to changes in forest cover in the Congo Basin caused by climatic changes over several million years.

The guenons all belong to the genus *Cercopithecus*, which means "long-tailed monkeys." Identifying guenons is often a matter of glimpsing a number of key identifying features—but this can be difficult when a troop is traveling through the canopy, backlit by the sky. No sooner do you fix your binoculars on one individual than it moves behind a clump of foliage or leaps into the next tree. A good field guide or knowledgeable companion is essential.

If a monkey's white throat and long red tail are visible through the foliage, and you glimpse a white nose, then you are watching Red-tailed Guenons. Red-tails are medium-sized monkeys living in primary and secondary forests across the Congo Basin and extending as far east as Uganda and western Kenya. They are not the only monkeys with a white nose: there are the Putty-nosed and Lesser Spot-nosed Guenons for a start. Nor are they the only ones with red on the tail: the closely related Mustached Guenon, which has a horizontal white bar on the upper lip, also has red on the lower two-thirds of its tail.

Mixed parties

To complicate matters further, Red-tailed Guenons often spend time with other monkeys; this behavior is known as polyspecific association. Glimpses of different features may therefore relate to different individuals from separate species. In the east, for example, Red-tails mix with Red Colobus, Blue Monkeys, Gray-cheeked Mangabeys and sometimes black-and-white colobus. Moreover, in some forests, hybrids between Red-tails and the larger Blue Monkeys can be seen. The hybrids have some features from both parents, but overall they are more like Blue Monkeys in appearance, although with a less obvious color. In the west of the Democratic Republic of Congo (DRC), Red-tails in swamp forests mix with Allen's Swamp Monkeys. In drier forests, they associate with Mona Monkeys (another guenon species), Black Mangabeys and Angolan Colobus.

Facing page This female Diana Monkey is pictured in captivity. In the wild in West Africa, females of this species often lead mixed-species groups of guenons and colobus monkeys.

There is surprisingly little dietary overlap between these mixed troops, and studies indicate that it is more often other species that join the Red-tails, not the other way round. This may be because Red-tails are good lookouts and are often the first to raise the alarm if a predator is spotted. In other species associations, the roles are often clearer; for example, Putty-nosed and Mustached Guenons often mix because the latter feed and keep watch on the ground, while the former feed in the canopy and watch the skies.

In many ways, Red-tails are typical guenons—their fur is tawny when seen close-up, and their manner is alert and curious. They live in troops of up to three dozen individuals, led by a dominant male with several females and young. Their diet is varied but dominated by fruit; they spend about a third of their day searching for insects in the vines at lower levels of the forest, half the day foraging in the middle levels and the rest in the upper canopy. Grooming is important, and rest periods are spent carefully parting the thick fur—either their own or a friend's—and picking out any bits of dirt, vegetation or external parasites.

Infanticide

Maturing male Red-tails leave their natal group, while females stay at home and inherit the same territory as their mother. A troop leader might be alpha male for as long as five or six years, but when he is deposed, the new leader may kill any babies still being suckled. Such infanticide is an instinctive behavior seen in many species and is explained by kin selection: as unnatural as this behavior may initially appear, there is an evolutionary advantage for a male to sire and raise his own young, rather than spend his time and energy in protecting another male's offspring. In addition, a female cannot usually conceive while producing milk, because stimulation of the nipple by the feeding infant sets off hormone reactions in her body that suppress the release of the next egg, thereby acting as a natural birth control. By killing the other male's infant, the new alpha male causes the female to become fertile more quickly and be receptive to his overtures. Males with the gene that triggers this behavior are likely to father more young, so evolution will in the long term favor infanticide.

Right This young Golden Guenon (*Cercopithecus kandti*) is just beginning to develop the beautiful orange-gold coat of the adult.

Tourists on their way to meet Mountain Gorillas are sometimes delighted to encounter another, smaller nonhuman primate while picking their way through the dense bamboo thickets. High-pitched chirrups and descending trills announce these monkeys' approach, and swaying bamboo fronds reveal medium-sized, solid-looking animals with dark limbs and a striking golden-red coat. These monkeys are Golden Guenons (*Cercopithecus kandti*), which are considered by some to belong to the Blue Monkey superspecies, *Cercopithecus mitis*, one of the most widespread types of African monkey.

The astonishing diversity of Blue Monkeys, guenons which are found from Sudan south to South Africa and Angola, appears to be linked to their dietary flexibility—they can adapt their diet to fit whatever is available. As forests retreated during drier periods, some populations of Blue Monkeys became isolated long enough for new species to evolve. If these populations developed sufficient differences, they did not interbreed successfully when the forests regrew, so they could be considered separate species. The Blue Monkey cluster of species (or subspecies, according to some authorities) are known as "greater periphery guenons" because they are found around the edge of the central Congo Basin, but not much in it. They include *C. albogularis*, the Sykes Monkey of Kenya, and *C. nictitans*, the Putty-nosed Monkey of Cameroon.

Isolated populations

Golden Guenons represent one such isolated population, found only in the cool, moist bamboo forests in the Virunga Volcanoes on the border between Rwanda, the DRC and Uganda and on the Rwandan side of Lake Kivu. They eat less fruit and more leaves than their close relatives the Blue Monkeys of Kibale Forest, Uganda, located only 125 miles (200 km) to the north. Studies show that Blue Monkeys eat parts of 59 plant species, compared to just 33 for Golden Guenons. The lower number for Golden Guenons is at least partly a result of their habitat: relatively few tropical fruit-bearing plants can cope with the cool temperatures in a montane habitat, so Golden Guenons have fewer fruits to choose from. The lower diversity of plant life at higher altitudes leads to a lower diversity of the animal species that eat the plants. Like Giant Pandas, they have specialized in eating bamboo, though they are also fond of the sweet yellow flowers of Saint-John's-wort trees.

Group size in Golden Guenons varies but averages around 30 individuals. In Rwanda and Uganda, some troops are now habituated to humans. The troop can be seen among a swaying sea of bamboo as the monkeys reach for young leaves and stuff them into their mouths. Golden Guenons are endangered, despite better protection in recent years. They have lost most of their habitat outside protected areas. Poachers kill them for their skins, and they sometimes get caught in wire snares set for antelopes.

PRIMATE PROFILE
GUENONS

SIZE Head and body length: 13–28 in. (33–70 cm); weight: 4–26 lb. (1.8–12 kg); males twice the size of females

APPEARANCE Short coat (color varies according to species); facial markings; round head; long body, hind limbs and tail

HABITAT Wide variety of montane and lowland forests, including wet and dry types; scrub

DIET Fruit, leaves, shoots, flowers, seeds, gums and exudates, insects, small animals

LIFE HISTORY Gestation: about 4-5 months; sexual maturity: male 6 years, female 4–5 years; life span: up to 20 years

BEHAVIOR Active in day; mainly tree-living; able to swim; forms groups of various types; scent marking in some species

In Kenya's smallest national park, the Saiwa Swamp, near Kitale, there is a good chance of seeing one of the most extravagantly marked monkeys in the world: De Brazza's Monkey. This guenon spends most of its time in the lower levels of the forest, only descending to the ground for about one-fifth of the day. Despite its ornate markings, it can be difficult to spot in the trees of this tiny patch of riverine forest, because it will sometimes stay in the same place without moving for hours. This behavior protects the attractive species from poachers, but confounds the efforts of conservationists trying to estimate its numbers. The small Kenyan population of De Brazza's Monkey is threatened with hunting and habitat loss, but it is right on the eastern edge of a pan-African distribution—the species is found in 11 countries from western Kenya and Ethiopia south to Angola and west to Cameroon—so as a whole it is not endangered.

Dealing with predators

De Brazza's Monkey is the only guenon that seems to be monogamous, although not in every population: multifemale groups led by a dominant male have also been seen. It is also one of the few guenons not to form multispecies groups. Males are almost twice the weight of females and announce their presence with a booming vocalization, enhanced by inflating their throat sac. They also scent-mark from their chest gland and give a yawning display, exaggerated by their long white beard and revealing their impressively long canine teeth. They show great courage in the face of predators, displaying and shaking branches, and may even rush forward to counter-attack.

The Diana Monkey is also beautifully marked, and also defends its territory with intergroup calling, backed up by aggressive encounters if necessary. The females play an important role by giving their intergroup call, a chattering cry, and this stimulates the male to give his *"hack"* call. Females form close ties and stay together in the center of the group, while the male spends much of his time on the periphery.

Diana Monkeys are found only in remaining patches of rainforest in West Africa, from Sierra Leone to Ghana. They share their forest with Olive Colobus and Lesser Spot-nosed and Campbell's guenons, and often form associations with them. Red colobus monkeys will also seek proximity to the territorial Diana Monkeys if they hear the calls of Chimpanzees, which regard red colobus as a favorite food.

Left The ostentatious markings of De Brazza's Monkey include a bright orange brow and white muzzle, beard and thigh-stripe.

Above This Blue Monkey is licking leaves. Forest primates gain much of their water from eating or licking wet vegetation, which reduces the need to drink from pools or streams where predators may lurk.

Diana Monkeys respond to alarm calls of other animals, such as birds, squirrels and duikers (forest antelopes); like Vervets, Dianas have different alarm calls for Leopards and eagles.

In 1977, a monkey skin from Salongo, in what was then Zaire (now the Democratic Republic of Congo), was described as a new species, thought to be a close relative of the Diana Monkey. This specimen is now known to be a juvenile Dryas Guenon, a little-known species that resembles a more dramatically marked vervet but has not yet been studied. There is clearly much left to discover about the exciting, still-evolving guenons and the important role they play in the ecology of Africa's forests.

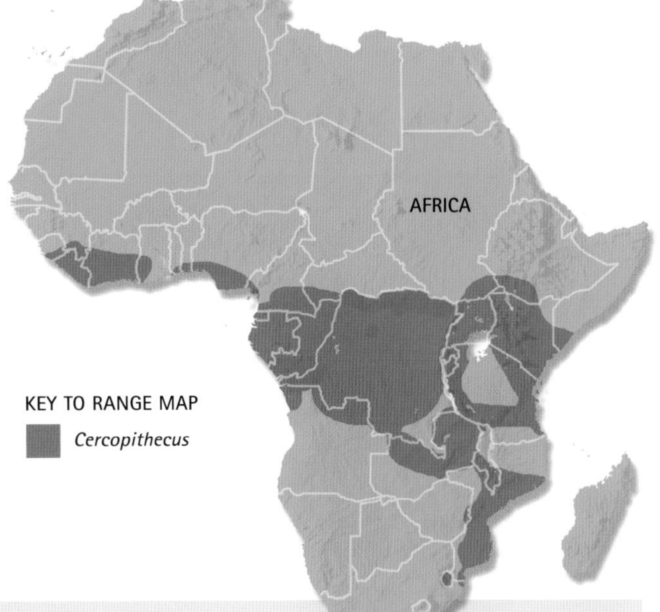

MAIN GUENON SPECIES

Scientific name	Common name	Where they live	Red List	CITES
Cercopithecus ascanius	Red-tailed Guenon	West to East Africa (Congo✪, Kenya✪, Rwanda✪, Tanzania✪, Uganda✪)	LC	II
Cercopithecus cephus	Mustached Guenon	Cameroon✪ to Angola (Gabon✪)	LC	II
Cercopithecus diana	Diana Monkey	Sierra Leone to Ghana✪ (Ivory Coast✪)	EN	I
Cercopithecus hamlyni	Owl-faced Guenon	Congo, Rwanda✪	NT	II
Cercopithecus l'hoesti	L'Hoest's Monkey	Burundi, Congo, Rwanda✪, Uganda✪	NT	II
Cercopithecus kandti	Golden Guenon	Eastern DRC, Rwanda✪, Uganda✪	EN	II
Cercopithecus mitis	Blue Monkey	Angola, Congo✪, Ethiopia to South Africa✪ (Kenya✪, Rwanda✪)	LC	II
Cercopithecus mona	Mona Monkey	Ghana✪ to Cameroon (Nigeria✪)	LC	II
Cercopithecus neglectus	De Brazza's Monkey	Angola, Cameroon✪ to Kenya✪	LC	II

RED LIST: CR = Critically Endangered EN = Endangered VU = Vulnerable NT = Near Threatened LC = Least Concern DD = Data Deficient NE = Not Evaluated ✪ = Best place to watch

Black-and-white Colobus

Watching a troop of black-and-white colobus leaping between trees is an unforgettable sight. Holding their arms outstretched, with long white fur cape flying and huge fluffy tail forming an aerial rudder, they look almost as though they should be able to fly. Unfortunately, they don't so much fly as plummet, crashing into the branches and clinging on tightly. Troop members tend to follow the same route, so if you see one jump, wait because others will likely follow.

The word "colobus" is derived from the Greek *kolobus*, meaning "mutilated one," and refers to the colobines' thumb, which is little more than a short stub. The fingers are long, forming flexible hooks that enable quick movement through the trees. The monkeys' lightweight bones also help to make them nimble leapers. Another important adaptation of colobines is their multichambered stomach, which enables them to exploit perhaps the most abundant food item in forests—leaves.

African colobines are less diverse than the Asian "leaf-monkeys," a reflection of the lower diversity in general found in African forests. They are the most arboreal of African monkeys and run on all fours along familiar routes through the branches. Traditionally there were thought to be two genera, *Colobus* for black-and-white forms and *Piliocolobus* for red forms, but many taxonomists now recognize the Olive Colobus (see pages 134–135) as a genus of its own, *Procolobus*.

Silky fur

Colobus species have beautiful long, silky fur, the black-and-white coloration of which breaks up their outline in the shifting shadows of the forest. All have very long tails, which in some species are tipped with a big white brush. Not all have the famous capes of long white hair: some just have white shoulder patches, and one, *Colobus satanas*, is all black. Black-and-white colobus spend most of their time high in the canopy, coming down to the ground occasionally. They live in a variety of lowland and montane forests, including rainforest and riverine savanna forest.

BLACK-AND-WHITE COLOBUS SPECIES

Scientific name	Common name	Where they live	Red List	CITES
Colobus angolensis	Angolan Colobus	Angola, Burundi, DRC, Kenya✪, Rwanda✪, Tanzania✪, Zambia	LC	II
Colobus guereza	Black-and-white Colobus or Guereza	Nigeria to Ethiopia, Kenya✪, Uganda✪, Tanzania✪	LC	II
Colobus polykomos	King Colobus	Gambia✪ to Ivory Coast✪	NT	II
Colobus satanas	Black Colobus	Cameroon✪, Eq. Guinea✪, Gabon✪	VU	II
Colobus vellerosus	Geoffroy's Colobus	Ivory Coast✪ to Nigeria✪ (Ghana✪)	VU	II

RED LIST: CR = Critically Endangered EN = Endangered VU = Vulnerable NT = Near Threatened LC = Least Concern DD = Data Deficient NE = Not Evaluated ✪ = Best place to watch

Below With their long, flowing cape and tail hairs, black-and-white colobus, such as this Mantled Guereza, were once a target for the fur trade.

Colobines have a specialized digestive system for dealing with leaves; toxic chemicals are neutralized in the forestomach before being fermented by bacteria in the main stomach. Foliage is not, however, the only thing they eat. Black-and-white colobus diets vary and may include fruit, flowers and even lichen—a food eaten by very few primates. In West Africa, the diet of the Black Colobus monkey may be made up primarily of seeds.

Social organization

Colobine social systems are highly diverse, ranging from small groups containing only a few individuals to huge groups with hundreds of members. Even within each species there is variation. Groups of Angolan Colobus may be as small as 10 in some areas, whereas in Nyungwe Forest, Rwanda, they form troops several hundred strong. Groups may consist of many females and their offspring, with the addition of one, two or more males.

Most colobines give birth to young that differ in color from their parents, and in all but the Black Colobus, the offspring are a milky white color. Births occur year-round, although peaks may occur at certain times; the King Colobus is an exception in having a specific birth season. Many females in the group help to care for the young while it retains its juvenile coloration. At about three months old, the young monkey will take on the color of the adults and lose its appeal to other females.

The familiar threats of habitat clearance and hunting continue to threaten all African colobus monkeys. As well as being slaughtered for the bushmeat trade, they are killed for their fur. Both native people and westerners wear the coats of these primates. One form of red colobus, *Piliocolobus badius waldronae*, was hunted to near-extinction by the middle of the 20th century.

PRIMATE PROFILE
BLACK-AND-WHITE COLOBUS

SIZE Head and body length: 19–27 in. (48–68 cm); weight: 17–30 lb. (7.4–13.5 kg); males larger than females

APPEARANCE Long black-and-white fur, pattern varies between species; slender body; long tail; small thumb

HABITAT Wide variety of forest types

DIET Leaves, fruit, seeds, shoots, flowers, lichen

LIFE HISTORY Gestation: 5–6 months; sexual maturity: male 6 years, female 4 years; life span: up to 30 years

BEHAVIOR Active during the day; tree-living; small groups with variable structure; grooming; associates with other species of monkey; loud calls to mark territory

Red Colobus and Olive Colobus

The name of the red colobus belies its true colors. From Gambia in the far west to Zanzibar off the coast of East Africa, the fur coats of these slender colobines differ according to region, with varying amounts of black, white, gray and brown, as well as rusty red. The most extreme form is the Udzungwa Red Colobus from Tanzania, in which the only red is a shocking spiky hairdo that contrasts with black-and-white body fur.

A clear-cut taxonomy of red colobus monkeys has yet to be resolved. Some authorities recognize only two species—the Red and the Olive Colobus—while many scientists now place the Olive Colobus in its own genus, *Procolobus*, and recognize up to nine species of red colobus in the genus *Piliocolobus*. Whether or not it should be placed in its own genus, the Olive Colobus differs from the red, and in more than just color.

Olive Colobus are seldom seen, leading secretive lives in the lowland moist tropical forests of West Africa, from Sierra Leone to Nigeria. They are active during the day and spend most of their time in the thick vegetation of the lower to middle levels of the forest, often near water. Red colobus species, on the other hand, live in the upper to middle canopy, in forests right across Africa. They also occur in lowland rainforest, but are equally at home in coastal, riverside or montane rainforest.

PRIMATE PROFILE
RED COLOBUS

SIZE Head and body length: 18½–27 in. (47–69 cm); weight: 15–23 lb. (7–10.5 kg); males larger than females

APPEARANCE Coat color and pattern varies with species, but usually has areas of red-brown fur; very long tail

HABITAT Primary and secondary forest; dry deciduous forest; wooded savanna; scrub

DIET Leaves, fruit, buds, shoots, fungi

LIFE HISTORY Gestation: 6½ months; sexual maturity: male 4 years, female 3–4 years; life span: not known

BEHAVIOR Multimale, multifemale groups, selectively feeding on tips or young leaves

Below With its long tail acting as an aerial rudder, and limbs spread like a skydiver, a Zanzibar Red Colobus leaps effortlessly between trees in the Jozani forest, Zanzibar.

Facing page A male Olive Colobus relaxes on a branch and carefully observes the forest. These monkeys hide in the trees and are rarely seen.

Red colobus monkeys usually live in large groups of up to 80 individuals, each containing many males, females and offspring. Olive Colobus groups are often smaller, typically with fewer than 12 members, including one or two males, several females and their offspring. Membership is fairly flexible, however, and individuals of both sexes may switch groups at some point in their lives. Olive Colobus also frequently forage around guenons (see pages 126–131), which may be a way of increasing vigilance for spotting predators.

Chimpanzee danger

Both the Olive Colobus and various red colobus species are hunted not only by humans but also by Chimpanzees, who share much of their forest home. Red colobus monkeys in particular provide an important source of protein for Chimpanzees, making up about 90 percent of the mammal prey of some populations. This can have a huge effect on the red colobus populations, and group size and composition may also change in response to the pressure of Chimpanzee predation.

Of the African colobines, Olive Colobus have the highest proportion of leaves in their diet—up to 85 percent. However, fruit, seeds and flowers also provide important nutrition. Red colobus also eat mainly leaves but also eat more fruit, seeds, flowers and fungi. Colobines also eat meals of clay, termite mounds and charcoal. These apparently unappetizing supplements help to neutralize the toxic compounds in the leaves that form a large part of their diet.

Unlike Asian colobines, female Olive and red colobus (and the Angolan black-and-white species) flaunt their sexual receptivity with a swelling and brightening of the skin around their sexual organs. The sex organs of all young red colobus monkeys resemble those of the adult female. This is thought to prevent adult males from mistaking the young for other males, which would result in them being evicted from the group. Infants are born to red colobus throughout the year, whereas Olive Colobus adhere to a strict birth season. Care of young by females other than the mother is rare in both species. Uniquely among the Old World primates, Olive Colobus mothers carry their young in their mouth. This could be because the infant, with its lack of a real thumb, cannot grip the short fur of its mother while traveling through thick undergrowth.

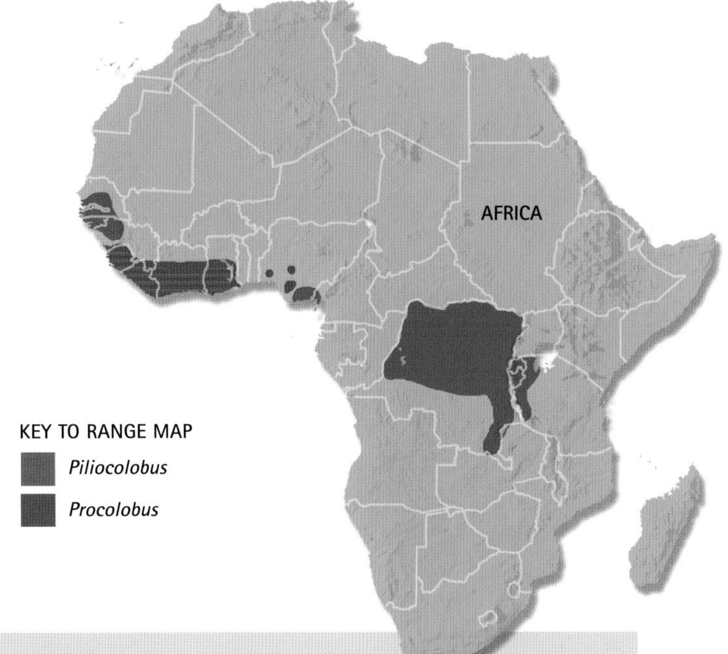

RED AND OLIVE COLOBUS SPECIES

Scientific name	Common name	Where they live	Red List	CITES
Piliocolobus badius	Western Red Colobus	Senegal to Ghana✪ (Gambia✪)	EN	II
Piliocolobus foai	Central African Red Colobus	Congo✪ to southern Sudan	DD	II
Piliocolobus gordonorum	Udzungwa Red Colobus	Tanzania	VU	II
Piliocolobus kirkii	Zanzibar Red Colobus	Zanzibar Island✪ (Tanzania)	EN	II
Piliocolobus pennantii	Pennant's Colobus	Congo✪, Eq. Guinea, Nigeria	EN	II
Piliocolobus preussi	Preuss's Red Colobus	Cameroon, Nigeria✪	EN	II
Piliocolobus rufomitratus	Tana River Red Colobus	Kenya✪	CR	I
Piliocolobus tephrosceles	Ugandan Red Colobus	Burundi, Rwanda, Tanzania, Uganda✪	NE	II
Piliocolobus tholloni	Thollon's Red Colobus	DRC	NE	II
Procolobus verus	Olive Colobus	Sierra Leone to Ghana✪	NT	II

RED LIST: CR = Critically Endangered EN = Endangered VU = Vulnerable NT = Near Threatened LC = Least Concern DD = Data Deficient NE = Not Evaluated ✪ = Best place to watch

Langurs

The various langurs and their close relatives the odd-nosed colobines (*see* pages 140–143) make up the Asian branch of the colobine family. Primarily leaf-eaters, langurs possess the leaf-adapted stomach characteristic of all colobines. They are held sacred by many people in Asia, but hunting remains one of the biggest threats facing them, and some now teeter on the brink of extinction.

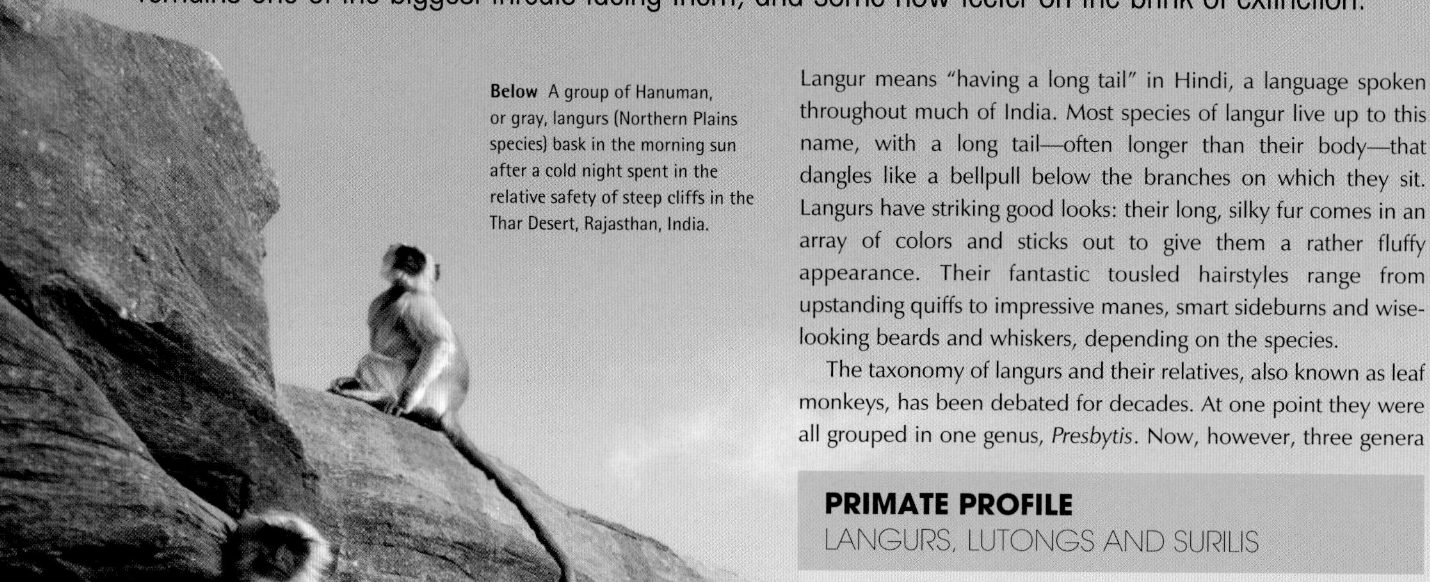

Below A group of Hanuman, or gray, langurs (Northern Plains species) bask in the morning sun after a cold night spent in the relative safety of steep cliffs in the Thar Desert, Rajasthan, India.

Langur means "having a long tail" in Hindi, a language spoken throughout much of India. Most species of langur live up to this name, with a long tail—often longer than their body—that dangles like a bellpull below the branches on which they sit. Langurs have striking good looks: their long, silky fur comes in an array of colors and sticks out to give them a rather fluffy appearance. Their fantastic tousled hairstyles range from upstanding quiffs to impressive manes, smart sideburns and wise-looking beards and whiskers, depending on the species.

The taxonomy of langurs and their relatives, also known as leaf monkeys, has been debated for decades. At one point they were all grouped in one genus, *Presbytis*. Now, however, three genera

PRIMATE PROFILE
LANGURS, LUTONGS AND SURILIS

SIZE Head and body length: 16–33 in. (41–84 cm); weight: 12–46 lb. (5.5–21 kg); considerable variation between species

APPEARANCE Coat color varies according to species; long tail; enlarged salivary glands

HABITAT Wide range of forest types; savanna, scrub, cliffs; cultivated land and urban areas

DIET Leaves, seeds, fruit, shoots, flowers, bark, stems, fungi, small animals

LIFE HISTORY Gestation: about 6 months; sexual maturity: male not known, female 3.5 years; life span: up to 33 years

BEHAVIOR Ground- and tree-living; loping gait; social groups vary according to species; infanticide follows takeover by new dominant male

Above This Thomas's Langur is sitting in the rainforest canopy of Sumatra, Indonesia. Males of this species may live in an all-male group or in a group with several females and one dominant male. In certain circumstances, males may kill the offspring of other males.

of these primates are widely accepted by primatologists, and the species list keeps growing longer as genetic studies reveal more diversity. Surilis remain in the *Presbytis* genus, and lutongs have been given their own genus, *Trachypithecus*. Lutongs and surilis share the forests of Indonesia and Malaysia; lutongs have also spread further north, as far as Vietnam. Meanwhile, the Hanuman langur was until recently considered a single species with several subspecies found across southern and western Asia, but these have now been reclassified as seven different species in their own genus, *Semnopithecus*, also known as gray langurs. Confusingly, lutongs and surilis are sometimes also referred to as langurs.

Langurs usually live in forests, many of which are tropical or subtropical, but there are some exceptions. Amazingly, some François's Lutongs live on vertical cliff faces, transferring their agility in the trees to mountains and sleeping in caves at night. Hanuman langurs are one of the most familiar types of monkey in the Indian subcontinent and are as much at home in busy city streets as in the forests.

Langur society

As a general rule, langurs live in groups containing one male, a few females and their offspring. There are, however, many exceptions to this rule. The Mentawai Langur, like the Pig-tailed Langur with which it shares its range (see page 139), is often seen in male-female pairs and may be monogamous. More research is needed to confirm this, because many "monogamous" primates have upon closer inspection turned out to have much more complex reproductive behaviors. In some other langur species there may be a few males in residence, but males are generally outnumbered by females.

Bachelor groups are also common, although this term can be misleading as the groups are not always all-male and sometimes contain young females. More research may reveal greater complexities within these groups. Males usually leave home (or are kicked out) and join bachelor groups, but it is not uncommon for females to switch groups during the course of their life, sometimes joining bachelors in the interim.

In Hanuman society, resident males do not enjoy power for long; they generally reign for about 18 months before other males begin to challenge them for prime position. The challenges may be quite violent and might be made by solitary males or by groups. If the dominant male is overthrown by a group of males, they will battle it out between themselves for the top spot. Infants of the predecessor may even be killed, perhaps to make females receptive sooner—mating often follows soon after the killing. Some species defend defined territories, with aggressive encounters occurring when groups meet at boundaries. Others are more tolerant of their neighbors.

Graceful movers

Langurs bound along with a graceful four-limbed locomotion to travel around their various habitats. Purple-faced Langurs tend to leap more than other species and if you are lucky you may see one make a spectacular drop of up to 50 ft. (15 m) through the forest branches. Although most langurs are primarily tree-living, Hanuman langurs are described as semiterrestrial, spending more time on the ground than other colobines.

Leaves provide important sustenance, and langurs, like other colobines, have stomachs containing different chambers to help digest plant material. Diets change with the seasons, though, and fruit and seeds are also important food sources. Flowers, buds and insects are also eaten, as well as less natural foods: urban monkeys eagerly accept handouts from their human cousins, and crop-raiding often provides easy meals.

Langur babies are often a different, brighter color than their parents. For example, the bright orange infant of the western subspecies of the Javan Ebony Langur (*Trachypithecus auratus mauritius*) contrasts sharply to its glossy black parents, while the gray-black Spectacled Leaf Monkey gives birth to yellow-orange infants. This may be a strategy to alert other group members to a newborn that requires greater protection, a reminder to be vigilant. The brightly colored youngster is groomed and carried by many females in the group. Typically only one young is born to each female, but births are often synchronized to occur at the same time. Young are usually weaned by the time they reach 12 months old, and the next follows up to two years later.

Left An orange-furred Delacour's Langur baby (center) is seen foraging with adults on rocks in their natural habitat, Van Long Nature Reserve, Vietnam.

Under threat

Some langur species are rapidly nearing extinction. Delacour's Langur, which lives only in Vietnam, is down to about 300 individuals. It is known locally as the *vooc mong trang*—"the langur with white trousers"—after the bold white markings on its rump and thighs. Population fragmentation and hunting for traditional medicine threaten to make this beautiful animal extinct by 2014.

Delacour's Langur is on the official list of the 25 most endangered primates, along with the Western Purple-faced Langur (*Trachypithecus vetulus nestor*) and the White-headed Langur subspecies known as the Golden-headed or Cat Ba Langur (*Trachypithecus poliocephalus poliocephalus*), after the only island on which it is found. In 2000, only 53 of the latter subspecies survived in the wild after suffering a shocking 98 per cent population decline in just 40 years. Again, hunting for traditional medicine is the biggest threat to its survival; various body parts are used to make a product known as "monkey balm." Its numbers are still critically low but are now slowly increasing (65 in 2007), due to the efforts of conservationists and more stringent poaching controls.

As the efficiency of hunting increases and the seemingly relentless habitat destruction continues to push these primates to the very limits of survival, the future seems bleak. But there is hope. Researchers and conservationists are making efforts to control poaching and protect forest habitats. Meanwhile, the beauty and charisma of langurs and odd-nosed monkeys increasingly appeal to primate watchers, and if ecotourism is managed correctly, it could bring significant conservation benefits.

Relationship with humans

Hindu folklore tells the story of Hanuman the monkey warrior, who burned his hands and face while trying to rescue Lord Rama's wife. This is how Hanuman langurs got their name—the langurs' black hands and face contrast sharply to their pale fur—and it explains the tolerance shown to city-dwelling monkey troops, even when they steal food. These monkeys can be seen on the streets, in buildings or sitting on rooftops gazing over the sprawling city below.

Given the noise and traffic, it seems surprising to see monkeys in an urban setting, but as cities encroach further and further into their forest homes, it is a case of adapt or die. India and China are rapidly developing economies, and in many areas their forests are being cleared at an alarming rate to make way for agriculture and other development. Not everyone in these regions regards monkeys as sacred, so as well as losing their natural habitat, these primates face a bigger threat—hunting. Across much of Asia they are hunted to be used in traditional medicine and for food; their skins are often used to make items such as drums; and they are also trapped and sold as pets.

MAIN LANGUR, LUTONG AND SURILI SPECIES

Scientific name	Common name	Where they live	Red List	CITES
Presbytis chrysomelas	Sarawak Surili	Borneo✪	DD	II
Presbytis comata	Javan Surili	Java (Indonesia)	EN	II
Presbytis femoralis	Banded Surili	Borneo, Indonesia✪, Peninsular Malaysia✪, Singapore✪	LC	II
Presbytis frontata	White-fronted Langur	Borneo	DD	II
Presbytis hosei	Hose's Langur	Borneo	VU	II
Presbytis melalophos	Sumatran Surili	Sumatra (Indonesia)✪	LC	II
Presbytis natunae	Natuna Islands Surili	Bunguran Island (Indonesia)	NE	II
Presbytis potenziani	Mentawai Langur	Mentawai Islands (Indonesia)✪	VU	I
Presbytis rubicunda	Maroon Leaf Monkey	Borneo✪, Karimata Islands (Indonesia)	LC	II
Presbytis siamensis	White-thighed Surili	Sumatra (Indonesia)✪	LC	II
Presbytis thomasi	Thomas's Langur	Sumatra (Indonesia)✪	LC	II
Semnopithecus ajax	Kashmir Gray Langur	India✪, Pakistan✪	LC	I
Semnopithecus dussumieri	Southern Plains Gray Langur	Western India✪	DD	I
Semnopithecus entellus	Northern Plains Gray Langur or Bengal Gray Langur	India✪, Pakistan✪	LC	I
Semnopithecus hypoleucos	Black-footed Gray Langur	India✪	DD	I
Semnopithecus priam	Tufted Gray Langur	India✪, Sri Lanka✪	VU	I
Semnopithecus schistaceus	Nepal Gray Langur	Nepal✪, Sikkim, Tibet	LR	I
Trachypithecus auratus	Javan Lutong	Indonesia (Java✪)	EN	II
Trachypithecus barbei	Tenasserium Lutong	Burma, Thailand	NE	II
Trachypithecus cristatus	Silver Leaf Monkey	Borneo✪, Burma to Vietnam (Thailand✪)	NE	II
Trachypithecus delacouri	Delacour's Langur	Vietnam	CR	II
Trachypithecus ebenus	Indochinese Black Lutong	Laos, Vietnam✪	DD	II
Trachypithecus francoisi	François's Lutong	China, Laos, Vietnam✪	VU	II
Trachypithecus geei	Gee's Golden Lutong	Bhutan, India✪	EN	I
Trachypithecus germaini	Indochinese Lutong	Burma to southern Vietnam (Thailand✪)	DD	II
Trachypithecus johnii	Nilgiri Langur	India✪	VU	II
Trachypithecus laotum	Laotian Langur	Laos	DD	II
Trachypithecus obscurus	Dusky or Spectacled Leaf Monkey	Bangladesh, Malaysia✪, Thailand✪	LC	II
Trachypithecus phayrei	Phayre's Leaf Monkey	Burma, China, Laos, Thailand, Vietnam	LC	II
Trachypithecus pileatus	Capped Langur	Bangladesh, Burma, China, India	EN	I
Trachypithecus poliocephalus	White-headed Langur	China, Vietnam✪	CR	II
Trachypithecus vetulus	Purple-faced Langur	Sri Lanka✪	EN	II

RED LIST: CR = Critically Endangered EN = Endangered VU = Vulnerable NT = Near Threatened LC = Least Concern DD = Data Deficient NE = Not Evaluated ✪ = Best place to watch

Odd-nosed Colobines

These Asian leaf-eating monkeys are known, for good reason, as the odd-nosed colobines. Similar in many ways to the closely related langurs, they differ in their extraordinary facial features. With the exception of the Proboscis Monkey (*see* pages 142–143), they have short, upturned noses. They also live in some extreme habitats.

Above This Red-shanked Douc langur is watching something intently, but in many primates curiosity must be tempered, because a full-face stare is also used and perceived as a threat.

Like their close relatives the langurs (*see* pages 136–139), odd-nosed colobines often possess long tails. An exception is the perhaps unfortunately named Pig-tailed Langur, which has a shorter, quite hairless tail and an upturned nose—peculiar features that recently prompted the creation of its very own genus. Until fairly recently, odd-nosed colobines were classified with the langurs. Scientists now generally agree that the odd-nosed colobines should be grouped separately, but relationships within this group tend to be more controversial. The chief disagreement concerns whether they should be further split into four subgroups (the doucs, Proboscis Monkey, snub-nosed monkeys and Pig-tailed Langur) or into just two (the doucs and the Proboscis Monkey).

Odd-nosed colobines are found in a variety of forest types. The Pig-tailed Langur of the Indonesian Mentawai Islands occurs in swamp or dipterocarp forests, while snub-nosed monkeys prefer the coniferous and broadleaved temperate forests of China and Vietnam, living at altitudes of up to 14,750 ft. (4,500 m), some of the highest primate habitats known. The visually striking doucs are most often encountered in evergreen and tropical forests, ranging from Vietnam through Laos to Cambodia. Also known as douc langurs, these monkeys were once thought to have a very limited distribution. However, research has shown them to be widely distributed, with habitats, ecology and social lives that vary between populations.

Food and feeding

Relatively little is known of what odd-nosed colobines eat, although their anatomy gives us a valuable clue. Like other colobines, these primates have stomachs divided into chambers suited to digesting plant material. Leaves are therefore likely to form a large part of their diet, but varying amounts of fruit, buds, flowers and bark may also be important foods in different seasons. Snub-nosed monkeys eat a lot of lichen, bark, buds and pine needles in winter, while the Black Snub-nosed Monkey (unusually for a primate) lives almost exclusively on lichen.

Odd-nosed colobines organize themselves socially in much the same way as langurs, with generally more females to a group than males, and one male often enjoying the harem to himself. Single-male groups of snub-nosed monkeys may share large home ranges of more than 8 square miles (20 sq km), traveling together in huge groups that sometimes contain as many as 600 individuals. In contrast to these vast gatherings, Pig-tailed Langurs prefer to spend their time in pairs and may therefore be monogamous. Pig-tailed Langurs also differ slightly in their use of sexual swellings to indicate sexual receptivity—they are the only Asian colobines known to use this strategy. Sex in all species is usually initiated by

PRIMATE PROFILE
ODD-NOSED COLOBINES

SIZE Head and body length: 16–33 in. (40–84 cm); weight: 12–46 lb. (5.5–21 kg); wide variation between species

APPEARANCE Coat color varies according to species; bare face; enlarged salivary glands

HABITAT Wide variety of lowland to montane forest types

DIET Leaves, seeds, flowers, shoots, bark, fungi, lichen, insects

LIFE HISTORY Gestation: 5–6 months; sexual maturity: 4–5 years; life span: 30 years

BEHAVIOR Active during the day; tree- and ground-living; variable social structure

females and is often concentrated at a particular time of year, so that young are born at around the same time. Single young are the norm, and females help to care for each others' young. Juvenile Black Snub-nosed Monkeys also help out on long journeys, by carrying their younger siblings and cousins.

The fur of the Golden Snub-nosed Monkey is still believed by some to prevent rheumatism, and such beliefs fuel the market for these primates in traditional medicine. Hunting is now leading to losses from which populations cannot recover, given the interbirth interval in odd-nosed monkeys of up to three years. The hunting pressure is made worse by large-scale habitat clearance to provide land and resources for a rapidly increasing human population, with the result that many of these extraordinary primates face extinction unless their habitat can be protected.

CHINA
SOUTHEAST ASIA

KEY TO RANGE MAP
Pygathrix
Rhinopithecus
Simias

Above Well-insulated against the cold, this endangered Golden Snub-nosed Monkey in China strips leaves at altitudes of up to 9,000 ft. (3,000 m) in summertime.

ODD-NOSED COLOBINE SPECIES

Scientific name	Common name	Where they live	Red List	CITES
Pygathrix cinerea	Gray-shanked Douc	Vietnam✪	EN	I
Pygathrix nemaeus	Red-shanked Douc	Cambodia, Laos, Vietnam✪	EN	I
Pygathrix nigripes	Black-shanked Douc	Cambodia, Vietnam✪	EN	I
Rhinopithecus avunculus	Tonkin Snub-nosed Monkey	Vietnam	CR	II
Rhinopithecus bieti	Black Snub-nosed Monkey	China	EN	II
Rhinopithecus brelichi	Gray Snub-nosed Monkey	China	EN	II
Rhinopithecus roxellana	Golden Snub-nosed Monkey	China✪	VU	II
Simias concolor	Pig-tailed Langur	Mentawai Islands (Indonesia)	EN	I

RED LIST: CR = Critically Endangered EN = Endangered VU = Vulnerable NT = Near Threatened LC = Least Concern DD = Data Deficient NE = Not Evaluated ✪ = Best place to watch

Proboscis Monkey

Take a dawn boat ride along a forested lowland river in Borneo and you might encounter the most bizarre-looking monkey on Earth stretching and sunning itself in the waterside trees. Proboscis Monkeys have huge pot bellies and the enormous fleshy nose of the adult male is so big that if he looks up quickly, it will flop back and smack him between the eyes.

Proboscis Monkeys are found only on the island of Borneo. Their pink face, pot belly and ambling gait, coupled with the males' massive nose, led local people in the Indonesian half of Borneo to call them "Dutchman monkeys," an unflattering reference to their former colonial rulers. The females lack the pendulous nose of the males, having instead a smaller, upturned nose. Proboscis Monkeys belong to the group of aptly named odd-nosed monkeys (see pages 140–141), which are in turn members of the Asian colobines, along with langurs. The taxonomy of the Asian colobines has been debated for decades.

Below Female and young Proboscis Monkeys lack the pendulous nose of the alpha male; this mother and young are sunning themselves in riverside trees as here in Sukau, Sabah, Borneo.

Above Male Proboscis Monkeys are easy to spot in Bako National Park, Malaysian Borneo; they make a "*kee-honk*" call through their extraordinary nose, which acts as a resonating device.

The Proboscis Monkey's body is mainly a reddish orange, but the legs, belly, tail and the rump patch where it joins the body are white-gray. Juveniles begin life with a blue face, but this becomes the adult pink at the age of three.

With the growth of ecotourism, Proboscis Monkeys have achieved celebrity status and receive a steady flow of intrigued visitors, all hoping to see the extraordinary males with their enormous nose. Proboscis Monkeys now rival orangutans as the most popular primates to watch in Borneo, and to save toiling through swamps, they are best viewed, hurtling and honking through the branches overhead, from the comfort of a *klotok* (riverboat) or canoe.

Swamp forest monkeys

Proboscis monkeys live in waterside forests—mangrove and *nipa* palm swamps near the coast or riverside forests further inland. To cross narrow rivers or inlets, they sway branches back and forth, building momentum to catapult them over the water, flying through the air and crashing into the trees on the other side. If the gap is too large, or they are taken by surprise, they make spectacular dives from the treetops, hitting the water with a loud bellyflop.

Once in the water, Proboscis Monkeys are adept swimmers, aided by partially webbed hands and feet. To escape danger, they can swim underwater for up to 65 ft. (20 m). A slow monkey

Above This infant has the characteristic blue face of Proboscis Monkeys when young. Proboscis infants, in common with all primate infants, are completely dependent on their mother for food, warmth, protection and transport.

may be snapped up by a lurking crocodilian, especially the False Gharial, so groups quickly swim across in single file, and hurry up a tree on the other side. They clamber and walk with a slightly ambling, four-limbed gait, sometimes rising to stand or waddle on two legs. Although Proboscis Monkeys have the standard colobine chambered stomachs to aid with leaf digestion, they eat other foods too and switch to dry, bitter fruit when it becomes available.

A Proboscis Monkey family contains one male and a few females (nonbreeding males form bachelor groups). Group size ranges from 3 to 23, but they are not territorial and may meet up with other groups throughout the day to feed or travel together or at night to share nearby sleeping trees. Meetings are generally amicable, although males may display and vocalize at each other. Males make a loud *"kee honk"* sound that may be enhanced by the nose acting as an organ of resonance. Whether for sound or looks, females seem to prefer males with larger noses and compete with each other to mate with the resident male, encouraging him by lip pouting, head-shaking and presenting. He seems to need no encouragement, however, for he has a permanently erect penis. The function of this peculiar trait is still a mystery.

Coloration

Births normally occur at night, and usually only one blue-faced infant is born, with sparse, dark fur. It will start to develop the red coloring of its parents at around eight months; the legs and lower arms turn gray later on. Daughters usually remain in the groups in which they are born. Young male Proboscis Monkeys leave home at around 18 months and join bachelor groups until they are mature enough to lead a harem of females on their own.

This species is endangered, and although the wetland areas in which the monkeys live have recently gained more protection, forest fragmentation restricts their movement between smaller forest patches. Habitat clearance for palm oil plantations poses a major threat, but the Proboscis Monkey's absurd appearance may be its saving grace. Its popularity with tourists generates an income for local people and strengthens the case for its conservation.

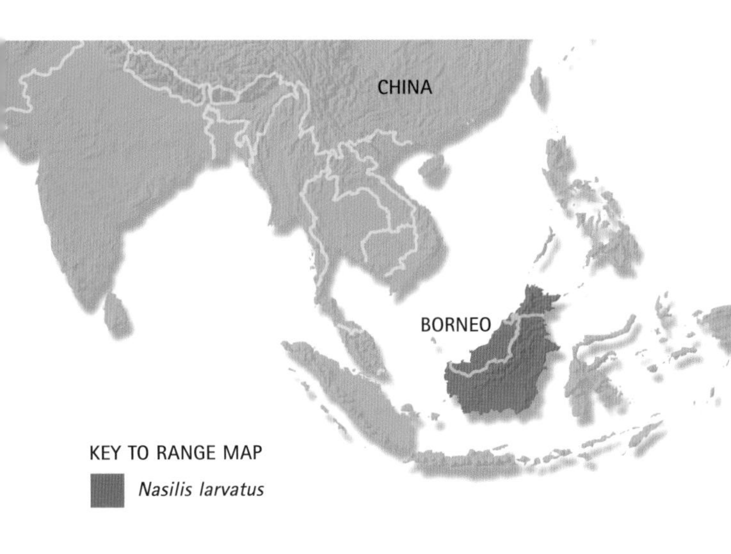

KEY TO RANGE MAP

Nasilis larvatus

PRIMATE PROFILE
PROBOSICIS MONKEY
Nasalis larvatus

SIZE Head and body: 24–30 in. (61–76 cm); weight: male average 22–47 lb. (10–21 kg)

APPEARANCE Orange-colored coat, darker on crown, gray limbs and belly; pink facial skin (adults); long, pendulous nose (adult males)

HABITAT Coastal *Nipa* palm, mangrove swamp, and lowland, riverine and peat swamp forest below 800 ft. (250 m)

DIET Leaves, seeds, fruit, flowers, animal prey

LIFE HISTORY Gestation: 5 months; sexual maturity: male 5–7 years, female 3–5 years; life span: up to 13.5 years

BEHAVIOR Active during the day; tree-living; swims; group-living; makes distinctive vocalizations through the nose (males)

RED LIST EN **CITES** I

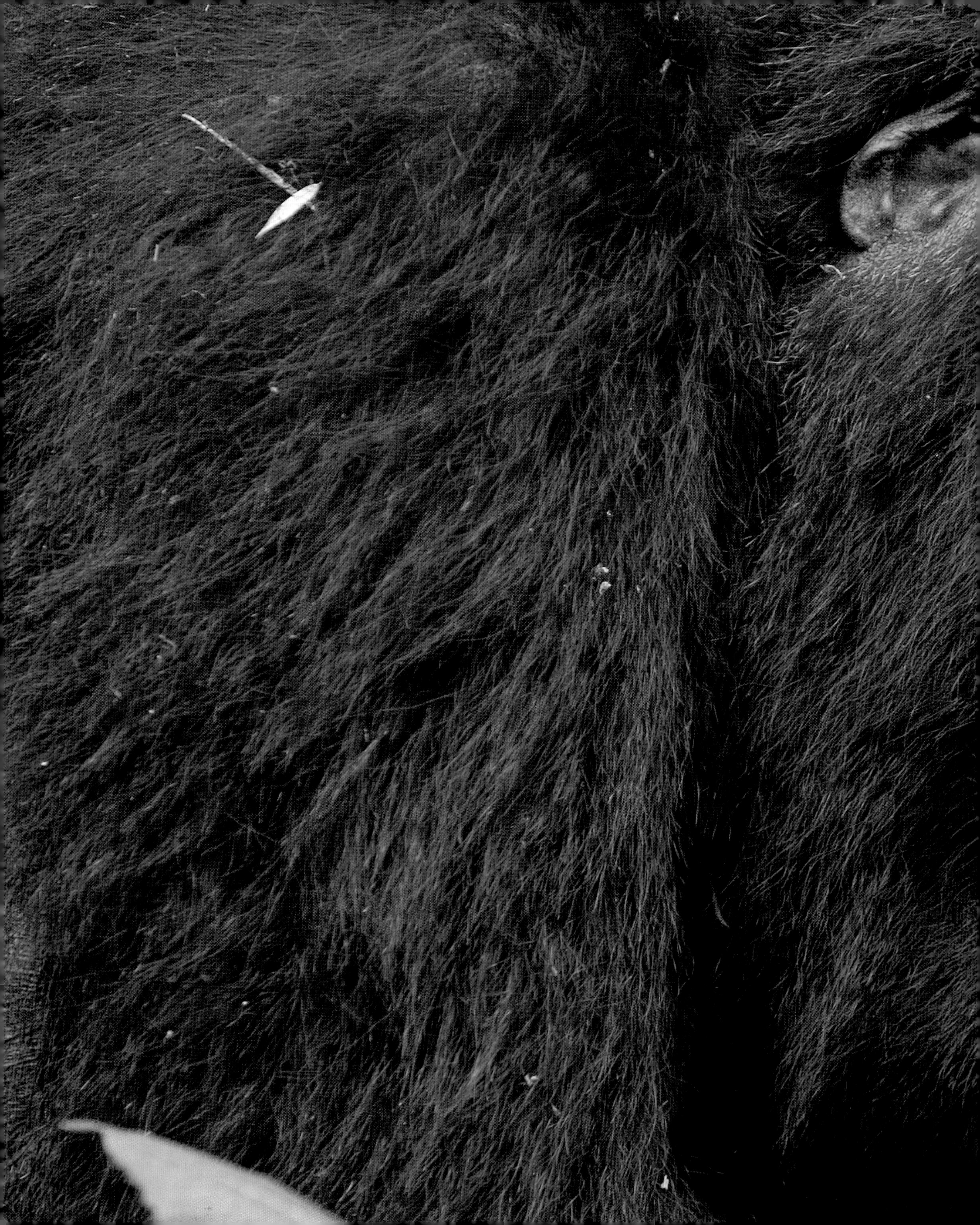

Apes

Main image Anyone lucky enough to exchange a glance with an ape, such as this Mountain Gorilla mother with her child, will realize that there is a quizzical mind at work behind those big brown eyes.

Introduction: Apes

Planet of the Apes, Tarzan of the Apes, King Kong—there are so many references to apes in popular culture that you would think everyone could recognize one. And yet there is widespread confusion about the differences between monkeys and apes. This is most obvious in languages in which the verbal distinction is unclear. The French word *singe,* for example, is usually translated as "monkey," but the term for great apes is *grands singes,* or "large monkeys." What then are the identifying characteristics of apes, in any language?

Large size, intelligence and the lack of a tail are three of the most obvious features of apes, which is why taxonomists consider humans to belong to the ape superfamily—the Hominoidea. The apes are further divided into the great and lesser apes. There are about a dozen lesser-ape species, including the gibbons and Siamang, and six species of nonhuman great ape—two gorillas, two orangutans, and the Chimpanzee and Bonobo. Great apes are unmistakable due to their size alone, but the same is not true of lesser apes because some macaques and baboons are larger than gibbons. A more reliable difference between apes and monkeys is the shape of the ribcage and length of the arms, both of which are adaptations for swinging beneath branches (brachiation). Monkeys have a ribcage that is flattened from side to side, with shoulder blades on the sides, which is suited to walking on all fours on branches; apes are barrel-chested, with shoulder blades on the back, giving more mobility to the arm for brachiation.

Above The coat color of the White-handed Gibbon of Southeast Asia may be cream, brown, black or red, but the hands and face-ring are always white. These apes, the fastest and most agile primates, swing vertically and horizontally from branch to branch and can jump midair gaps of up to 40 ft. (12 m) in the forest.

The relative length of arms and legs is called the intermembral index. Apart from humans, whose long legs for walking upright give an index of 72, all apes have arms longer than their legs (giving an index of more than 100), while the reverse is the case in all monkeys, in which arm length is equal to or less than leg length (producing an index of 100 or less). Siamangs have an intermembral index of 147—the highest of all primates—because their gangly arms are nearly one and a half times the length of their legs. Gorillas, with an index of 116, have the longest arms of the African apes; they do brachiate in play and sometimes when feeding in trees, but spend most of the time on the ground, where they walk on the soles of their feet and knuckles of their hands.

Senses and intelligence

Apes use vision more than their sense of smell and therefore have a flattened face with a reduced snout. The bony ring around the eyes is pronounced, serving to protect the eyes and also to frame the face, which helps to convey information through different expressions. Ears are relatively small and immobile in apes, but sound is nevertheless important for communication, and a wide range of vocalizations are used.

Apes are known for their intelligence, but this is a more difficult feature to measure objectively. One indication of intelligence is brain size, which can be estimated by measuring the cranial capacity—the size of the brain case in the skull. Scientists compare the volume in skulls of different species by filling the empty skull with small beads, then pouring them into a measuring cylinder. Apes have larger brains relative to their body size than other primates, and humans have the largest of all. Research into ape cognition—how apes use their large brains—is one of the most fascinating areas of primatology. Much of this work has been carried out in captivity, where experiments are easier to control than in the wild, although detailed field studies are now revealing how their high level of intelligence helps apes to thrive in their natural environment. Intelligence brings adaptability and an ability to reason; for example, some apes make and use tools to gain access to foods that would otherwise be unavailable to them.

Within the apes there are many different systems of social organization. These include the largely solitary existence of orangutans, the monogamous family groups of gibbons and many human cultures, the female-dominated communities of Bonobos, the male-dominated communities of Chimpanzees and the polygamous extended family groups of gorillas and some humans. The number of individuals in any social group is influenced by the quality, distribution and quantity of food resources in the territory in which the group ranges.

Since apes are largely fruit-eating animals, they have evolved the ability to carry a mental map and remember patterns of fruiting, so their movements around their habitat are not random but are based on a relatively systematic search for food. Some species of ape also eat meat, and it is in the cooperative hunting of monkeys by Chimpanzees that we see the closest echoes of human hunting behavior.

Facing page The trusting gaze of this young silverback Mountain Gorilla is testament to his confidence in humans; but having won his trust, it is essential that we protect him from poachers.

Above A Bonobo makes eye contact at Lola ya Bonobo sanctuary near Kinshasa, DRC. The orphans in this community have been rescued from the illegal pet trade. Plans are underway to return them to the wild within the species' former range.

Left A young Agile Gibbon plays with his father now that he is confident enough to make forays away from his mother's embrace.

Gibbons and Siamang

The most evocative sound heard throughout the forests of Southeast Asia is the whooping, wailing song of gibbons. These long-limbed, fast-swinging apes perform their beautiful song, a duet between male and female, for two main reasons—it reinforces the bond between the singers and it announces to their neighbors that this is their territory. The vocalizations are different in each species, so they may also act as an aid to identification.

Gibbons are found in the canopy of forests from Bangladesh south to Borneo and east as far as Vietnam and China. There are four groups of species that vary in size from about 11 lb. (5 kg) to 33 lb. (15 kg), and in color from black to brown or buff. Some species have different-colored males, females and infants, while others show color variants within the same sex. The number of species has been the subject of much debate, with some authors recognizing as many as 16, and others as few as 9 but with numerous subspecies. New genetic research, however, shows that the molecular distances between the four basic kinds of gibbon are at least as great as those between humans and Chimpanzees, so they should each be considered a separate genus.

Monogamous pairs

The classic image of gibbon society is that of a monogamous couple, singing together to greet the dawn, and bringing up their single infant in around 100 acres (40 ha) of fruit-rich forest. The interval between births is normally two to four years, and young mature after six to nine years, so a well-established family group is likely to consist of the two parents plus two or three young. By the time a fourth infant is born, the eldest will have left the territory. This resemblance to a human nuclear family has made gibbons an attractive role model. But more detailed field studies of White-handed Gibbons and Siamangs have shown that all is not what it seems—these apparently faithful pairs have been observed mating with the neighbors during intergroup encounters. In one study, copulations outside the pair amounted to 9 percent of matings observed, and some gibbon groups have been seen to accept extra adults, at least on a temporary basis. It is not yet known whether these flirtations result in conception, but clearly gibbon society is more complex than was thought.

Right The Siamang is the largest species of gibbon; unlike great apes, gibbons have two pads of thick, horny skin, called ischial callosities, to protect their bottom when sitting on a branch.

PRIMATE PROFILE
GIBBONS AND SIAMANG

SIZE Head and body length: 18–35 in. (45–90 cm); weight: 10–33 lb. (4.5–15 kg); siamangs are larger than gibbons

APPEARANCE Coat color varies according to species and sex

HABITAT Deciduous monsoon and evergreen forest

DIET Fruit, leaves, flowers, insects

LIFE HISTORY Gestation: 7 months; sexual maturity: male 6.5 years, female 9 years; life span: up to 44 years

BEHAVIOR Active during the day; small family groups based around monogamous pairs; highly territorial; variety of vocalizations

Why then, do most gibbons live in pairs? Various explanations have been offered, but it seems most likely that these fruit-eating canopy dwellers can take best advantage of their limited food supply by operating in pairs. Although leaves—and flowers when available—make up about a third of a gibbon's diet (more in the case of Siamangs and Black Gibbons), and insects are a vital supplement, gibbons eat much more fruit than anything else. A large group of gibbons would quickly deplete the fruit in its home area, and therefore the advantages of being in a big group would be outweighed by the disadvantages of needing to feed so many.

The favorite gibbon feeding technique is to hang from a branch by one hand while gathering fruit and putting it in the mouth with the other. Their relatively small body size enables gibbons to reach the ends of branches, and this niche is where gibbons excel. Studies of gibbon ecology reveal that they play an important role as seed-dispersal agents. One study in Borneo found that they disperse the seeds of 81 percent of the species they eat, and germination trials show that many of these species germinate better after passage through the gut. The predilection of gibbons for figs makes them particularly important for the dispersal of the seeds of these trees, which are important for thousands of insects, birds and other animals, including humans. It follows that the future of biodiverse forests in Southeast Asia is closely tied to the survival of gibbons. Unfortunately, these primates are among the most endangered species on the planet. Urgent action is needed to conserve the various species of gibbon and their forest homes.

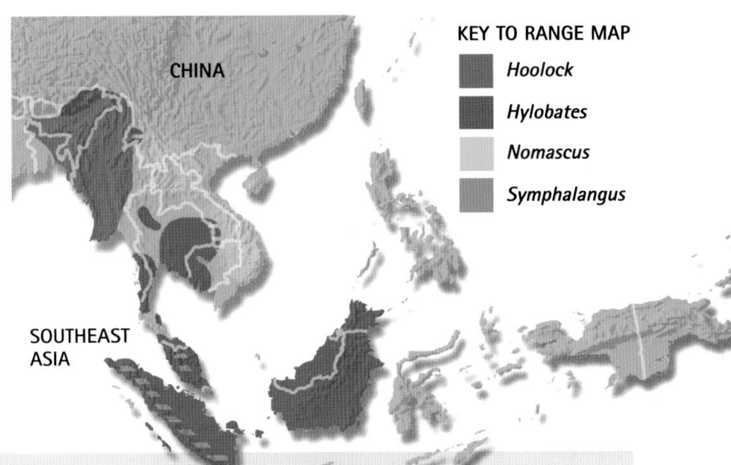

GIBBON AND SIAMANG SPECIES

Scientific name	Common name	Where they live	Red List	CITES
Hoolock hoolock	Hoolock or White-browed Gibbon	Bangladesh, Burma✪, China, India	EN	I
Hylobates agilis	Agile or Dark-handed Gibbon	Peninsular Malaysia✪, Sumatra (Indonesia)✪	LC	I
Hylobates klossii	Kloss's Gibbon	Mentawai Islands (Indonesia)	VU	I
Hylobates lar	White-handed or Lar Gibbon	Burma, China, Laos, Malaysia✪, Sumatra (Indonesia)✪, Thailand✪	LC	I
Hylobates moloch	Javan Gibbon	Java (Indonesia)✪	CR	I
Hylobates muelleri	Mueller's or Bornean Gray Gibbon	Borneo✪	LC	I
Hylobates pileatus	Pileated or Capped Gibbon	Cambodia, Thailand✪	VU	I
Nomascus concolor	Black Gibbon	China✪, Laos, Vietnam✪	EN	I
Nomascus gabriellae	Golden-cheeked Gibbon	Cambodia, Laos, Vietnam✪	VU	I
Nomascus hainanus	Hainan Gibbon	Hainan Island (China)✪	CR	I
Nomascus leucogenys	Chinese White-cheeked Gibbon	China, Vietnam✪	DD	I
Symphalangus syndactylus	Siamang	Malaysia✪, Sumatra (Indonesia)✪	LC	I

RED LIST: CR = Critically Endangered EN = Endangered VU = Vulnerable NT = Near Threatened LC = Least Concern DD = Data Deficient NE = Not Evaluated ✪ = Best place to watch

Brachiation

Watch children on a jungle gym in a park, marvel at trapeze artists in a circus, notice how people hold the overhead straps on a crowded bus—it is clear that humans, like all apes, are good at swinging. But brachiation—swinging from hand to hand as a means of locomotion—is quite difficult for us. With our long, heavy legs, our arms soon tire. Gibbons, on the other hand, have evolved to be brachiators par excellence and are the finest aerial acrobats of all mammals.

Above With their elongated fingers and extra-mobile joints, gibbons have perfected the art of brachiation. Note how this White-handed Gibbon keeps his opposable thumb beside his fingers, forming a quick-grab and release hook.

In the Tarzan books, American author Edgar Rice Burroughs fantasized that the young Tarzan learned from apes how to swing through the "middle terraces" of the African rainforest. But since those books were written in the 1920s, field scientists have discovered that while African apes do climb trees, they do so only to find fruit or build nests, and when they need to reach the next tree, they usually climb down and knuckle-walk on the ground. When the Tarzan books were made into films, even the athletic actors chosen for the lead role found themselves unable to live up to Burrough's fantasy of a human brachiating through the trees, so the device of conveniently placed vines with no roots was invented, actor and vine together forming a swinging pendulum.

Adaptations for brachiating

Gibbons use a form of pendulum motion but suspend themselves from an arm rather than a vine. Anatomical adaptations to achieve this include long fingers that easily hook over a branch and a short thumb that makes this action easier; when their long arms are raised, their center of gravity is low in the body, and this is helped by their short legs, so that the torso acts as the bob, or weight, of a pendulum. This maximizes the momentum of an arm swing: the gibbon lets go on the upward curve, using kinetic energy to travel through the air while reaching out with the other arm. In the forest canopy, the next branch might be at any angle, so a gibbon's wrist must be able to twist to match it. The wrist bones in gibbons have evolved to allow a much greater degree of rotation than in other primates, to the point where the wrist joint has virtually become another ball-and-socket joint like that in the shoulder: a gibbon can spin around beneath a branch almost without its hand moving.

Due to these adaptations, when gibbons need to walk—usually along branches too thick to hang beneath, but seldom on the ground—they do so on their legs, holding their long arms out above their head for balance. It is for this reason that some early naturalists believed gibbons might be the closest living relatives to humans, being the only other primates habitually to walk upright on two legs. Once other evidence was taken into account, however, it became apparent that gibbons branched off first from the great ape–human line. Molecular studies now put this date at approximately 18 million years ago.

Gibbons are the only apes to sit on branches when sleeping rather than build nests in the trees or on the ground; this is why gibbons are also the only apes to have ischial callosities, or sitting pads, on their bottom. These pads of thick, toughened skin protect the flesh. A gibbon infant must grip its mother tightly at all times, especially when she is swinging, because the youngster needs to resist the combined force of gravity plus centrifugal force. Mother gibbons often use their tucked-up legs to support their infant while swinging through the trees. But even adults can make mistakes. One study of 118 gibbon skeletons found that about a third of them had healed fractures, and some individuals had more than one. This suggests that falls do occur during the lifetime of a gibbon, but they are not necessarily fatal.

Above Heavy orangutans seldom brachiate as freely as gibbons, preferring instead that at least two limbs grasp a branch or vine for safety, as this Sumatran Orangutan mother demonstrates to her clinging infant.

Orangutans

Orangutans are the largest tree-living animals on Earth. Among the first apes to be shipped to Europe in the 18th century for display to an incredulous public, they nevertheless have been studied less than gorillas and Chimpanzees in the wild. The fascinating secret life of these apes is at last being revealed at several study sites in Sumatra and Borneo, but it is a race against time to understand orangutans and their needs before their last remaining forest is destroyed.

The name orangutan comes from the Malay words *orang*, meaning "person," and *hutan*, meaning "forest," and these "people of the forest" were known to the inhabitants of Borneo and Sumatra thousands of years before they were finally described by western science. Orangutans show differences in appearance between the sexes, to such an extent that early European naturalists thought that the massive males with their prominent cheek pads were a different species to the smaller, more elegant females. Adult males weigh up to 175 lb. (80 kg), twice the weight of a female, so living high in the canopy involves a certain degree of danger.

Due to their size, orangutans do not brachiate as freely as gibbons; instead they use a more measured clambering technique to travel through the trees, known as quadrumanual climbing. Orangutan hip joints allow great mobility, and big toes that are able to grasp enable their feet to grip as firmly as hands: usually, at least two limbs are holding on at any one time. Flexible woody plants are also used to good effect: orangutans swing on vines and sway back and forth on tall saplings, until an outstretched arm can catch hold of the next tree's branches, which are pulled nearer until the orangutan can transfer to a strong branch. For the human observer down below, negotiating slippery logs and swamps, keeping up with a moving orangutan can be quite a challenge, but any frustration is always mixed with sheer admiration.

Studying wild orangutans is made more difficult still by the remote locations of their remnant wild populations and by the fact that these primates spend most of their lives alone, or at least some distance from their nearest neighbor. Females have overlapping home ranges of 0.4–2 square miles (1–5 sq km).

PRIMATE PROFILE
ORANGUTANS

SIZE Head and body length: 45–54 in. (115–137 cm); weight: 90–200 lb. (40–90 kg)

APPEARANCE Red-brown shaggy fur; very long arms; cheek pads and throat sac (adult males); beard (both sexes)

HABITAT Tropical rainforest, peat-swamp forest

DIET Fruit, leaves, bark, flowers, birds' eggs, various small vertebrates; honey

LIFE HISTORY Gestation: over 8 months; sexual maturity: male 9.5 years, female 7 years; life span: up to 60 years

BEHAVIOR Tree-living; active during the day; solitary; alpha male long-calls to deter rivals and attract females; nests in trees

Left This Sumatran Orangutan has grasped a termite nest in its mouth and will eat the insects inside. The varied diet of this species includes more than 200 plant species, supplemented by termites and small mammals, including slow lorises.

Below Orangutans have the longest childhood of any nonhuman animal, spending their first eight years with their mother learning survival skills.

Dominant males, in contrast, range over a much greater area that includes the ranges of several females. None of these ranges are necessarily fixed, and orangutans of both sexes travel according to the availability of food.

When fruit is plentiful, orangutans eat virtually nothing else. Fruiting trees, however, are scattered and bear fruit at different times. Some species of tree produce huge quantities of fruit—a phenomenon known as mast fruiting. This enables orangutans to gorge themselves for a few days, quickly putting on weight to carry them through lean periods. Research indicates that orangutans may be able to build up a mental map of where fruit trees are and when they are likely to be in season. Overall, fruit makes up about 50–60 percent of their diet. Other food items include leaves, flowers, pith, fungi, honey, termites and small mammals.

During their long childhood, young orangutans study botany and food preparation by watching their mother closely and by sampling her food choices—different parts of different plants are edible, and inedible parts are discarded. Adult orangutans are very skilled at processing food; if you stand beneath an orangutan feeding in the forest, be prepared for a barrage of fruit husks, twigs and other detritus to rain down. These provide rich pickings for terrestrial animals, such as pigs and rodents.

Tool use

Orangutans in Suaq Balimbing, Sumatra, are remarkable in being the only ones known to regularly make and use tools to extract honey or termites from inside tree trunks. They strip the leaves off a stick and shorten it to about 1 ft. (30 cm) long, if necessary, then fray the ends; next, holding onto the trunk with hands and feet, one end of the stick is held in the mouth and the other is pushed into the bees' or termites' nest. The stick is turned around so that the honey- or termite-covered tip can be sucked while repeating the probing of the nest with the other end. This kind of tool use has never been observed in Borneo, but all orangutans use some tools; the use of large leaves as rain hats or umbrellas, for example, has been noted at all study sites.

Sumatran Orangutans differ from their Bornean relatives in a number of other ways. Until recently they were considered to be forms of the same species, but studies of their DNA have since proved that their evolutionary lines separated between one million and two million years ago, and as a result the pair were upgraded from subspecies to full species. Physically, Sumatran and Bornean orangutans can be told apart by the lighter orange-cinnamon coloration and more slender build of the Sumatran species, which also has more hair on the face and, in males, flatter cheek pads. Sumatran Orangutans spend more time socializing than the Bornean species, which may be a result of the greater abundance of fig trees in Sumatra—this provides sufficient fruit for several individuals to eat in the same location.

Slow reproduction

Male orangutans reach sexual maturity in their midteens, but they are careful not to challenge the local dominant male, who is easily recognized by his enlarged cheek pads, or flanges, and his powerful "long call"—a booming "*uuuuh*" sound interspersed with guttural gurgles and grunts. Young males may try to force females to copulate with them, but when a female is receptive she prefers to seek out the dominant male. He will consort with her for a couple of days, guarding her from other males and even fighting with any that challenge his status. It was originally thought that this behavior would result in the dominant male fathering all of the infants in his home range, but DNA paternity studies showed that 6 out of 10 infants were actually fathered by the smaller males without cheek pads, so clearly dominance isn't everything.

A single infant is born to the female after a gestation of eight-and-a-half months, and it must immediately cling firmly to its mother's body hair as she clambers through the canopy. The infant is totally dependent on its mother, experimenting with solid food at only about 11 months. Suckling continues for five or six years; a juvenile may stay with its mother even after she gives

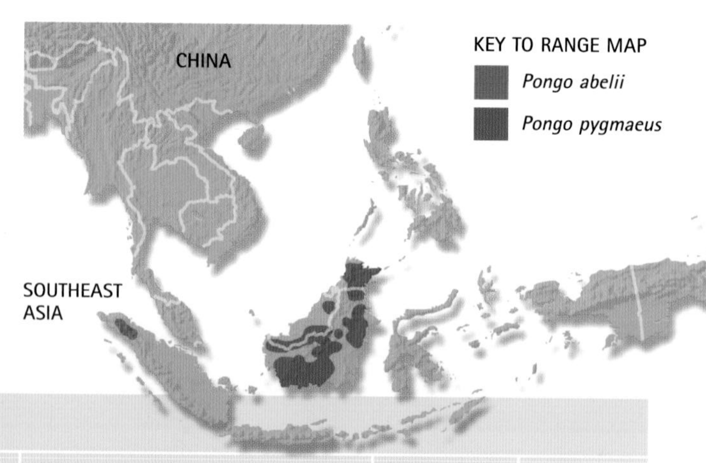

ORANGUTAN SPECIES

Scientific name	Common name	Where they live	Red List	CITES
Pongo abelii	Sumatran Orangutan	Sumatra (Indonesia)✪	CR	I
Pongo pygmaeus	Bornean Orangutan	Borneo✪	EN	I

RED LIST: CR = Critically Endangered EN = Endangered VU = Vulnerable NT = Near Threatened LC = Least Concern DD = Data Deficient NE = Not Evaluated ✪ = Best place to watch

Below This young male Sumatran Orangutan is just beginning to develop the cheek pads, or flanges, that characterize the adult male.

Right This dominant male Bornean Orangutan has fully developed cheek pads and throat sac, both of which may enhance the resonance and power of his long call.

birth to her next infant. The interbirth interval is the longest of any species of primate, at eight or nine years. This makes orangutans very vulnerable to hunting pressure: if more than 1–2 percent of an area's orangutans are killed every year, the population will soon start to decline and will eventually be destroyed.

Humans have been hunting orangutans for a long time. Cooked orangutan bones from 35,000 years ago have been excavated from caves in Sarawak, and some forest tribes in Borneo still occasionally eat ape meat. Fortunately, this practice has never reached the level of commercial bushmeat hunting in Africa. But it is likely that the mainland orangutan, which was widespread across Southeast Asia as recently as 10,000 years ago, disappeared as a direct result of human persecution. At current rates of deforestation and hunting, the Sumatran Orangutan may follow the mainland species into extinction within 20 years, and the Bornean species soon after that. Only when conserving rainforests for the long term is seen to be beneficial will orangutans enjoy a more secure future.

Gorillas

Gorillas are among the most spectacular animals on the planet. For centuries, travelers recounted tales of humanlike monsters in Africa, but it was only in 1847 that an American missionary brought back a skull and the species was described by western science. Ever since, gorillas have fascinated the public. At first it was their huge size and dramatic displays of size and strength when faced by big-game hunters that riveted people. But when field studies revealed a gentler side to gorilla family life, this proved equally compelling.

Gorillas live in the tropical forests of Central Africa, on either side of the Congo Basin, from near sea level in the west to altitudes of nearly 13,000 ft. (4,000 m) in the east. People across this region have long respected the humanlike qualities of gorillas but feared their strength and supposed ferocity. These conflicting views led some tribes to regard gorillas as totem animals, almost like neighbors, while others perhaps only a few miles away believed that by eating gorillas they could gain some of their strength and power. As a result, gorillas were—and still are—hunted for their meat and for use in traditional African medicine.

The first gorilla specimens to be collected by explorers were shot in what is now Gabon. In 1862, a French-American hunter, Paul du Chaillu, wrote a popular book about his adventures hunting gorillas. His descriptions of facing "a hellish dream creature" charging, standing up and beating its chest in rage were widely copied, and decades later his story provided the inspiration for the film *King Kong*, about a giant gorilla-like ape.

The heaviest primate

A typical silverback male is as tall as an average human, but about three times heavier and with much longer arms. Females are smaller, being less than half the weight of an adult male. The silver saddle of fur that gives the silverback his name is a badge of maturity that males develop when they reach puberty; before that, they are known as black-backs because, like the females, they have black hair all over, except for the fingers and toes, palms of the hands, soles of the feet, chest and face. In adult females, the chest has sparse hair except for the breasts, which are hairless. In males, the development of chest hair is, curiously, the reverse of that in human males (who are hairless when young and develop a hairy chest after puberty). Immature male gorillas have hairy chests until puberty, at about 10–12 years, when the whole of the front of the chest loses its hair, revealing dark gray skin over the pectoral muscles. Adult males also have longer hair on the arms, which serves to exaggerate their size when they stand with arms wide in a four-limbed strut. Western Gorillas often have a crown of reddish hair, sometimes extending onto the nape and even occasionally the shoulders. Their body hair is shorter and sleeker than that of Eastern Gorillas.

How many species?

Eastern Gorillas were not discovered by western science until 1902, when they were described as a new species, the Mountain Gorilla. Later this classification was revised, and for most of the 20th century it was thought that there was a single species of gorilla with three subspecies—the Western Lowland, Eastern Lowland and Mountain gorillas. However, when DNA from different populations was eventually analyzed, it became clear that the major difference was not between gorillas living in mountain and lowland areas, but between the eastern and western varieties. It was agreed that these forms were so distinctive that they were redefined as separate species. The most northerly outlying population of gorillas, on the Nigeria-Cameroon border, was also found to have differerent characteristics to other Western Gorillas and is now believed to be a subspecies, known as the Cross River Gorilla. With fewer than 300 individuals left, living in several sub-populations occasionally linked by migration, this is the world's most endangered kind of great ape.

Opposite The saddle of silvery fur, crest of bone and muscle, bare chest and long, shaggy arm hair of this Mountain Gorilla are secondary sexual characteristics that male gorillas develop at puberty, at about 11 years old. This individual lives in Rwanda's Volcanoes National Park.

Above Western Lowland Gorillas often have striking reddish brown hair on the crown, which sometimes extends down onto the shoulders.

Above It was long thought that gorillas hated water, until the discovery of swampy clearings called *bais* in Central Africa. In these pools, Western Gorillas wade in to feed on water plants, as seen here in Odzala National Park, Congo.

Western Lowland Gorillas are the most numerous gorilla subspecies, but their population has crashed, from an estimated 100,000 as recently as the mid-1990s to just half that today. This is a result of widespread logging, which has not only destroyed much of their rainforest home, but has also made this remote habitat accessible to the illegal commercial bushmeat trade. The rapid fall in gorilla numbers due to habitat loss and hunting has been exacerbated by outbreaks of Ebola virus, a disease that kills virtually all apes it infects, including humans. Efforts are now underway to halt this catastrophic decline (see pages 40–43).

Family life

Gorillas live in stable family groups led by a silverback, whose role is to protect the family from danger. The chief natural dangers are predators—mainly Leopards—and rival unrelated silverbacks. A group may contain several females with infants and juveniles, but usually only one silverback, unless it is a long-established group and the dominant male's sons have themselves become silverbacks. When males reach this stage they may become more and more peripheral, until they are confident enough to strike out on their own as lone silverbacks; sometimes these gorillas leave with a brother of similar age for company.

Lone silverbacks travel further than family groups, interacting with their neighbors and getting to know the terrain. The social interactions are stylized displays that involve a series of hoots

Left The Mountain Gorillas of the Virunga Mountains are distinguished by their longer, thicker fur that covers the brow ridge—this is an adaptation to the cold in their montane forest habitat.

PRIMATE PROFILE
GORILLAS
Gorilla beringei, Gorilla gorilla

SIZE Head and body length: up to 70 in. (180 cm); weight: 200–400 lb. (90–180 kg)

APPEARANCE Black or brownish gray coat, adult males develop a silver back and a crest of bone and muscle on the top of the head; stocky body with long arms; large nostrils; large hands

HABITAT Tropical forest (montane and swamp)

DIET Fruit, leaves, stems, flowers, seeds, bark, roots, some invertebrates

LIFE HISTORY Gestation: 8.5–9 months; sexual maturity: 8–10 years, usually longer in males; life span: up to 50 years

BEHAVIOR Active in daytime; groups of 5–60, including males and females; not territorial; makes new nest at night, males on ground, females and young on ground or in trees

building up to a crescendo, accompanied by a chest-beat and, as a finale, the ground is thumped or a branch may be torn off a tree. As the males gain experience, their displays of strength and self-confidence start to impress young females in the groups they visit, and eventually they may be joined by one or more females to start a new group. If an alpha male dies, a young silverback may attempt to step into his place, although he may not have the necessary qualities of leadership to hold the group together. In such cases, dissatisfied females will transfer to a group led by a more impressive silverback. The males left behind form a bachelor group and remain like this until one or more of them is able to attract new females to the group.

Females leave the group they were born in at around eight years of age and are likely to produce their first young a year or two later. The single infant (twins are rare) clings to its mother with hands and feet, at first on her belly; then, after about five months, it rides on her back, looking around as she knuckle-walks through the forest. Group members spread out to feed, avoiding competition for plants, and communicate using a quiet contact call known as a "BV," or belch vocalization (so named because early observers thought that the gorillas were simply burping). Infants are weaned at three to four years, during which time they learn which plants are edible and how to prepare them by watching their mother.

Left Many books state that gorillas don't climb trees because they are too heavy. In fact, they often do so, carefully, to build nests and find fruit and other foods. They also climb to obtain a better view. This gorilla is climbing to eat bark.

Right A silverback beats his chest with open hands on the slight hollow beneath the pectoral muscle, making a loud "*pok-pok*" sound that may be heard miles away.

Species differences

The most obvious visible difference between Mountain Gorillas and other gorilla subspecies is the hair. To cope with the cold, thin air in their high-altitude habitat, Mountain Gorillas have evolved thick, shaggy fur, as well as broader chests to improve their lung capacity. The gorillas spend their days foraging for leaves, stems, roots and occasional fruits in the montane rainforests that cloak the slopes of volcanoes at 8,000–13,000 ft. (2,400–4,000 m).

Like lowland gorillas, Mountain Gorillas form stable family groups, usually comprising several females and their young, led by a silverback. However, multimale groups are more common in Mountain Gorillas than in other gorilla subspecies. In most cases they are made up of a father and his sons, or several brothers. There can, nevertheless, be fierce rivalries, and silverbacks put on impressive displays of strength—hooting, pummeling their chests and ripping branches off trees. Although the aggression seldom escalates into physical violence, fights can result in serious wounds inflicted by long canine teeth and massive hands, so it is advantageous to both sides if a dispute can be settled by sizing up the opposition before the first blow is struck.

Dian Fossey's legacy

Several Mountain Gorilla groups in Rwanda have been studied since 1967, when Dian Fossey established the Karisoke Research Center, and their family histories are now known in impressive detail. There have been significant changes in group structure over the decades, and it may be that these changes are the result of human pressure on the 60 square miles (160 sq km) of Volcanoes National Park, which form part of the trinational Virunga Conservation Area, covering a total of 175 square miles (450 sq km). For example, in the 1970s, most gorilla groups contained 10 to 20 individuals, but over the years some groups have steadily grown in size. Now the largest are three times that size, with up to 60 individuals, including several silverbacks. Researchers can only speculate on the causes of this change, but every year brings new and exciting observations.

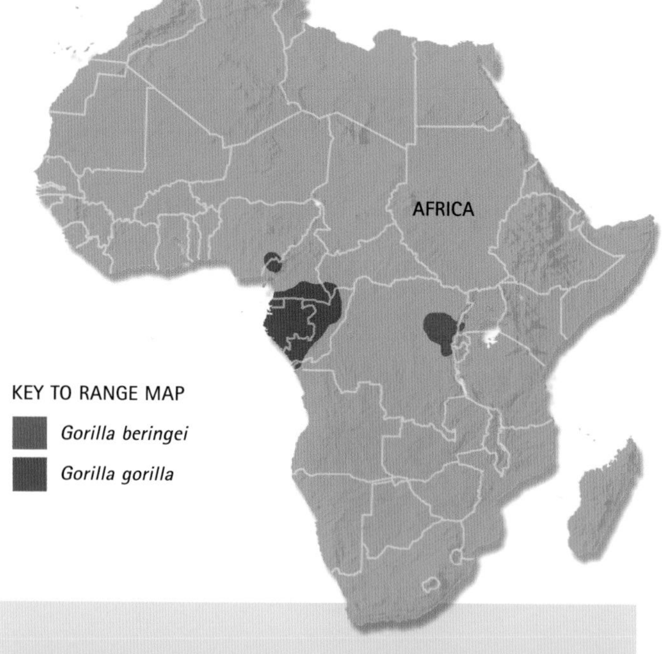

GORILLA SUBSPECIES

Scientific name	Common name	Where they live	Red List	CITES
Gorilla beringei beringei	Mountain Gorilla	DRC✪, Rwanda✪, Uganda✪	EN	I
Gorilla beringei graueri	Eastern Lowland Gorilla	DRC✪	EN	I
Gorilla gorilla gorilla	Western Lowland Gorilla	Angola, Cameroon, Congo✪, CAR✪, DRC, Eq. Guinea, Gabon✪	CR	I
Gorilla gorilla diehli	Cross River Gorilla	Cameroon/Nigeria border	CR	I

RED LIST: CR = Critically Endangered EN = Endangered VU = Vulnerable NT = Near Threatened LC = Least Concern DD = Data Deficient NE = Not Evaluated ✪ = Best place to watch

Above After a morning spent eating bamboo shoots, a nutritious seasonal food, this family of Mountain Gorillas is resting. On the left, two infants play, on the right, adults relax, all under the watchful eyes of the dominant silverback.

Since 1979, the methods originally developed to habituate these gorillas to the presence of primate researchers have been applied to other gorilla groups for tourists to visit. Several family groups have been able to tolerate a daily one-hour visit by small parties of excited humans, and there has been a dramatic improvement in human-gorilla relations in the region as a result. Gorilla-viewing safaris raise valuable income and create local jobs, leading to both governmental and community support for gorilla conservation, despite the intense pressure for land in one of Africa's most densely populated nations.

Thanks to the pioneering work of Dian Fossey and her colleagues, Mountain Gorillas have become an icon for the conservation movement. The film *Gorillas in the Mist* is a dramatization of Dian's book of the same name, and portrays a lone scientist battling against the odds to protect the charismatic study animals she has grown to love. The reality, of course, is far more complex, but the publicity generated by the film has certainly helped. Despite years of war, genocide, refugee crises and poverty in the surrounding human communities, Mountain Gorilla numbers are slowly increasing, as are tourism revenues. It is at best a fragile recovery, and there is always the risk that careless contact with a tourist carrying a virus might trigger an epidemic in the Mountain Gorilla population. Moreover, there are new threats on the horizon—climate change may affect rainfall patterns and global warming may modify the upland vegetation zones in Central Africa. So Mountain Gorillas still face an uncertain future, but they do seem to have a better chance of survival than most other populations of apes.

Below This silverback Mountain Gorilla calmly surveys his domain from Mount Visoke, looking east over the low-lying bamboo forest in Rwanda's Volcanoes National Park. Surrounded by food plants, Mountain Gorillas select the best parts to prepare each succulent mouthful.

Chimpanzee and Bonobo

Chimpanzees are often described as our closest living relative, but there are in fact two species sharing this distinction—the Chimpanzee (once called the Common Chimpanzee) and the Bonobo, or Pygmy Chimpanzee. Both share about 99 per cent of their DNA with humans, which means that genetically they have more in common with us than with gorillas or orangutans. Sometimes called the "Left bank" and "Right bank" Chimpanzees because of their distribution on either side of the Congo River, they display striking differences in social behavior, giving a valuable insight into the evolution of the species some see as the third Chimpanzee—humans.

"Oooooh-uuh, oooo-uuh, oooooo-uuh, ho-uh, ho-uh, ha-uh, ha-uh, ha-uh, ha-ha-ha-ha!" Chimpanzee pant-hoots ringing through the forest are immediately recognizable as an excited sound, often meaning that a member of the community has found a rich food supply, such as a fruiting fig tree. The callers not only convey their personal excitement but also invite others to come and join in the feast. Chimpanzees live in wide-ranging communities of up to 100 individuals, and they also exercise choice by joining up in different groups for different periods of hours or days, in what is known as a fission-fusion society.

Long-term studies in the field have revealed that Chimpanzee communities are a tangled web of family ties, long-lasting friendships and strategic—sometimes short-lived—political alliances. Males form cliques and jostle for social status, but they cooperate in joint endeavors such as hunting or patrolling the boundaries of their territory to see off strangers from neighboring communities. If such a border patrol comes across a lone stranger, it may attack and in some circumstances even kill—behavior that can seem chillingly reminiscent of tribal warfare in human societies. On the other hand, Chimpanzees can be tolerant and cooperative, and they have been seen to show kindness, compassion and concern for the welfare of others who might or might not be close relatives.

The Chimpanzee playground

Like us, Chimpanzees learn the rules of their society and gain experience through play during a long childhood. They remember favors, bear grudges and will gang up with allies to attack a more dominant individual who has, for example, bullied them. For a human observer, being in a forest surrounded by a Chimpanzee argument can be very intimidating: the aggressors *"WAAAAH"* loudly, victims scream in fear and onlookers join in vocally in support of one side or the other. Sometimes the melee escalates to physical violence as blows or bites are traded, but after things calm down appeasement generally follows—hands are extended to seek reassurance, there are plenty of breathy grunts and lip-smacking, and there may be embraces or sessions of social grooming.

These fascinating scenes are enacted in forests and woodland across the African tropics, from Senegal and Guinea in the west to Tanzania in the east. Chimpanzees are the most widespread great ape, currently being found in 21 or 22 countries.

Left These Eastern Chimpanzees, subspecies *Pan troglodytes schweinfurthii*, are gathered around an alpha male known as Frodo hoping for a tidbit as he feeds on a dead Bushbuck fawn, in Gombe National Park, Tanzania.

Below An infant Chimpanzee's world resounds with hoots and screams, scents and tactile sensations, all experienced first from the security of its mother's arms.

PRIMATE PROFILE

CHIMPANZEE AND BONOBO

Pan troglodytes, Pan paniscus

SIZE Head and body length: 29–38 in. (74–96 cm); weight: male 88–132 lb. (40–60 kg), female 70–104 lb. (32–47 kg)

APPEARANCE Black coat, naked face, infants have a white tail tuft. Chimpanzee: infant face is pink to brown and darkens with age; may have white beard. Bonobo: lighter weight and more slender build, longer limbs, black skin, red lips, hair has central parting on top

HABITAT Primary and secondary forest, dry woodland, wooded savanna, from lowlands to 10,000 ft. (3,000 m)

DIET Fruit, leaves, flowers, seeds, plant galls, animal prey (including mammals, birds, ants, termites, wasps, grubs); Chimpanzees hunt and eat other primates, especially colobus

LIFE HISTORY Gestation: about 8 months; sexual maturity: male 13 years, female 11 years; life span: up to 50–60 years

BEHAVIOR Active in daytime, makes sleeping nests in trees at dusk, uses many kinds of tools, large territorial communities dominated by males in Chimpanzees and by females in Bonobos; feeds in shifting parties; Bonobos have sex for social reasons as well as to reproduce

The question of whether Chimpanzees still survive in Burkina Faso has yet to be resolved. However, in Central Africa, south of the curve of the mighty Congo River, the same scenes of ape communities meeting and greeting take on a different tone. The excited hoots are an octave or two higher and include almost birdlike chirrups and squeaks. The females are the ones in charge, and when there are disputes, instead of violence there is sex. Welcome to the remarkable world of Bonobos.

Bonobos

The Bonobo, also called the Pygmy or Gracile Chimpanzee, was the last great ape species to be recognized by western science. It was first noted as different from Common Chimpanzees in 1929, then described as a distinct species in 1933. Genetic evidence indicates that Chimpanzees and Bonobos diverged between 1.7 and 2.7 million years ago. The most visible distinguishing features are the latter species' red lips, which contrast with the dark skin of its face, and the central parting of the hair on top of its head. Now famous for its "make love not war" system of social organization, the Bonobo remains the least-studied great ape. Ironically, the reason is that this peace-loving primate lives only in the war-torn DRC (formerly Zaire), making research extremely difficult.

Bonobo feeding parties tend to be larger than those of Chimpanzees and include more females than males. Both species eat meat, and Bonobos seem to prefer duikers (forest antelopes) to monkeys and hunt less often than some Chimpanzee communities. Three of the four current Bonobo study sites are in dense rainforest, while the fourth is on the southern edge of the species' range at Lukuru, where the forest opens out to savanna woodland. Here, Bonobos sometimes walk on two legs across grassland carrying food, looking rather like an artist's impression of Darwin's missing link. Indeed the Bonobo's femur is longer than in the Chimpanzee, and its thigh muscles are bigger, so the species does look more humanlike. The different ecology of

Below Orphaned by the bushmeat trade, sold as pets and confiscated by the Congolese authorities, these young Bonobos huddle together for comfort in the Lola ya Bonobo sanctuary near Kinshasa.

Above Bonobos are easily distinguished from Chimpanzees by their red lips, centrally parted hair on the crown and more slender build.

Bonobos at Lukuru may yet yield many more surprises. Given the variation in Chimpanzee behavior between different sites, it is likely that before long many of the behaviors thought not to occur in Bonobos will be seen in this species too.

The trade in Chimpanzees

In contrast with the more slender Bonobo, Chimpanzees are sometimes referred to as Robust Chimpanzees. They were once known as Common Chimpanzees, but that name is no longer appropriate. During the 20th century, their population declined from about two million to fewer than 200,000. They became extinct in four, perhaps five, countries and in many others they now survive in fragmented populations of a few dozen or a few hundred. The cause of this severe decline is twofold: habitat loss and fragmentation, through logging and the conversion of forest to agriculture; and hunting, both for bushmeat and to capture live infants for sale to zoos, pet shops and research laboratories.

The trade in Chimpanzees and Bonobos has been illegal since the 1970s. It goes on because unscrupulous traders offer cash to hunters with few other ways to make money. What seems a tempting sum to a village hunter is a tiny fraction of the price a wealthy buyer will pay for a young ape. Tragically, some collectors and trainers think more of their personal prestige or the money they can make than the fact that a family of apes will have been killed in order to capture a live infant. And infants often die during capture or transport, so each one that makes it alive to a competent carer represents perhaps 10 to 15 apes lost from the wild.

CHIMPANZEE SPECIES

Scientific name	Common name	Where they live	Red List	CITES
Pan paniscus	Bonobo or Pygmy Chimpanzee	DRC✪	EN	I
Pan troglodytes	Chimpanzee, Robust Chimpanzee or Common Chimpanzee	Angola, Burundi, Cameroon, CAR, Congo✪, DRC, Eq. Guinea, Gabon✪, Ghana, Guinea, Ivory Coast✪, Liberia, Nigeria, Rwanda✪, Senegal, Sierra Leone, Tanzania✪, Uganda✪	EN	I

RED LIST: CR = Critically Endangered EN = Endangered VU = Vulnerable NT = Near Threatened LC = Least Concern DD = Data Deficient NE = Not Evaluated ✪ = Best place to watch

The Cultured Ape

In 1960, a young Englishwoman traveled with her mother to set up a camp on the shores of Lake Tanganyika to study Chimpanzees. Jane Goodall was an unlikely candidate for the job, having few academic qualifications and no field experience, but something about her impressed her boss, the famed Kenyan anthropologist Dr. Louis Leakey. From this tentative beginning emerged the Gombe Stream Research Center in Tanzania—the longest continual study of any animal population—and the Jane Goodall Institute for Conservation and Research.

Above Watching an ape delicately crack a nut using a heavy stone tool can be unnerving. Here, two Bonobos observe another use skills originally taught by humans. Some Chimpanzee communities use tools naturally, others do not, and nor do wild Bonobos, so this is a good example of cultural transmission of skills.

One of the pivotal discoveries Goodall made in the early days was that Chimpanzees make and use tools—something scientists and anthropologists had always viewed as a uniquely human behavior. When she sent Leakey a telegram reporting her observations of tool use, he delightedly replied, "Now we must redefine 'tool,' redefine 'Man,' or accept chimpanzees as humans."

As more Chimpanzee field studies were carried out, it became apparent that the kind of tools they used varied from place to place. It was found that Gombe Chimpanzees used grass stems and thin wands as a probe to fish for termites, but a mere 125 miles (200 km) to the south, Japanese primatologist Toshisada Nishida found that Chimpanzees of the Mahale Mountains did not. They would, however, often use a probe to dip for ants and would then wipe the ants off against their hand to make it easier to eat them. At Gombe, the Chimpanzees were only occasionally seen ant-dipping, and they were never seen to use the wiping method for efficiently transferring the biting insects to the mouth. Moreover, young Chimpanzees at both sites were seen to closely observe their mother's feeding techniques and then try them for themselves, gradually improving with practice. Clearly, this was cultural transmission of learned behavior down the generations.

As more scientists spend time observing other Chimpanzee populations across Africa, the list of behaviors that might be termed "culture" lengthens. Not all of these learned behaviors involve tool use. For example, the Mahale Mountains Chimpanzees often clasp each other's hands over their heads while grooming, and yet this has never been seen in Gombe. The hand-clasping habit has been recorded in Kibale Forest in Uganda (although the position of the hands is different) but not in Budongo, also in Uganda.

TOOLS USED BY CHIMPANZEES

Chewing up leaves to make a spongy wadge and using it to drink rainwater from an inaccessible tree hole

Cracking nuts with rocks (in Ivory Coast and Guinea) or wood (only in Ivory Coast); this behavior has not yet been seen in Central or East Africa, but it has recently been reported in Cameroon

Pounding the soft center of a palm tree crown with the base of a palm leaf (seen only at Bossou in Guinea)

Using a leafy stick as a fly whisk

Making a sitting pad of large leaves

Using leaves as napkins to clean the body, to dab a wound to inspect bleeding, or to provide a clean surface to inspect or squash an external parasite, such as a louse

Using objects as weapons, including branches as clubs and pieces of wood or rock as aimed projectiles; most recently, in 2007, a female Chimpanzee in Senegal was seen using a sharp stick to spear a bushbaby in a tree hole (it is too early to say whether this is a cultural behavior or a clever individual that has just invented the technique)

In the late 1990s, data from all the main Chimpanzee study sites was pooled and a total of 39 cultural variants emerged. These were defined as behaviors that are "customary or habitual in at least one Chimpanzee community, yet absent without ecological explanation at another." Just as someone might recognize a person from his or her home town by an expression or gesture, so it became possible for primatologists to watch a video of Chimpanzees in a forest and deduce where it was filmed by the fact that certain learned behaviors feature in the film. Similar collaborative studies have been made of other great apes: 19 cultural variants have so far been noted in orangutans, for example.

It has become apparent that cultural variation did not emerge *after* humans evolved, which used to be the general consensus; instead, it was probably present much earlier in our family tree. The roots of culture would seem to date back to the common ancestor of all great apes and humans some 15 million years ago.

Below It takes years of practice for a Chimpanzee to master the art of termite fishing, but the methods are cultural because they are passed on from generation to generation, and they vary between Chimpanzee populations across Africa.

Humans

Human beings are extraordinarily variable in size and color, and yet our big brains and adaptations for walking on two legs make us anatomically distinct from other primates. This led 19th-century scientists to separate humans and great apes by placing our species in its own family, Hominidae. But when DNA studies revealed the closeness of apes and humans, the logical conclusion was that apes are hominids too. Some taxonomists now even argue that Chimpanzees and Bonobos (*see* pages 162–165) should share the genus *Homo* with us.

Humans are the most successful and widespread large mammal on Earth, although it remains to be seen whether the planet can survive the growing human biomass. This is a remarkable state of affairs, and looking back at our family tree can perhaps help us to understand our place in nature and how we reached this position of unrivaled dominance. Somewhere between 90 million and 50 million years ago our earliest primate ancestors, which resembled tree shrews, split from the other primitive mammals. At some point between then and now, many branches in the family tree split and expanded, some primate groups disappearing as climates changed and habitats altered, others adapting or clinging on in isolated forest refuges. The moment at which the human lineage separated from the other apes—maybe four million or five million years ago—is, quite literally, a bone of contention.

The Toumai mystery

Little is known of ancestral Chimpanzees or gorillas. Unfortunately for paleontologists trying to piece together our family tree, fossils do not often form in forests. So there was much excitement in 2001 when an apelike skull dating back 6–7 million years was found in Tchad in north-central Africa. Officially described as *Sahelanthropus tchadensis*, but nicknamed "Toumai," the species' characteristics are still a topic of heated debate. With a brain the size of a Chimpanzee but a flattened face more like that of modern humans, Toumai may be one of several things: the last common ancestor of Chimpanzees and humans; or an early gorilla ancestor; or a related ape that died out. In 2007, the picture was complicated further by the discovery, in Ethiopia, of a handful of teeth (but no bones) that look like gorilla teeth, and

which date back 10 million years—this is 3 million years earlier than DNA studies put the first separation of the gorilla lineage. Only the discovery of more fossils will clarify where these few controversial teeth fit into the wider scheme of things.

Human development

The molecular evidence suggests that our ancestors some 250,000 generations or 5 million years ago were the same as those of Chimpanzees and Bonobos. Since then, we have grown apart, developed more and more sophisticated tools, and learned to farm our favorite species of plants and animals. We have worked out how to create comfortable tropical microhabitats—both next to the skin through the use of clothes and around our family group through the ability to control fire in our homes—regardless of whether we live in a desert or an Arctic wasteland outside.

Variations in genes that control body size and skin and hair color have produced the diversity of ethnic races we see today. A great variety of traditions and cultures has also flourished across every continent. However, the human genome mapping project tells us clearly that humans are, genetically, a surprisingly homogenous bunch. Our liking for travel and sex with strangers means that gene flow has continued unabated. In the 160,000–195,000 years since our species emerged in Africa, and the 100,000 years since we spread out to colonize new continents, only very small changes in genetic makeup have produced our diversity of races.

Cultural change, however, occurs much more quickly than genetic evolution. Humans have a highly developed sense of cultural identity. We use music and songs, costumes and body adornment, dance and dialect to develop and display our social background. Whether this is in a tribal context, or in a football team, a religion, a business association or a teenage gang, knowledge of the group etiquette is essential for social acceptance and success. Language is at the heart of cultural identity—in New Guinea, for example, where more than 700 languages have been documented and intertribal communication is by pidgin English, the term for friend is *wontok*, meaning "one talk."

Throughout much of human history, small groups of related families defended their resources from strangers and tribal warfare was common. However, this powerful xenophobia is offset by the many cultures that have a tradition of welcoming strangers and the human fascination with things that are different. Human nature is essentially cooperative and our social relations are based on reciprocity—we form alliances, remember favors and, all too often, bear grudges against those who have wronged us.

In any group of people there will be some individuals with the urge to explore beyond the place in which they grew up. In the 21st-century world of global communications, traditional cultural barriers are breaking down, and while some regret the loss of diversity, few lament the better relations between people across the planet. Human history is a story of increasing the circle of trust, from family to tribe to nations, and is about a growing acceptance that all humans are worthy of respect. The question is: can this widening circle of trust and compassion now extend beyond our species to the rest of the primate family tree and beyond?

Below Humans are the most numerous primate, and we consider ourselves the most successful, but if our population growth does not level off, we will run out of resources, threatening all other primates on the planet in the process.

Where to Watch Primates

There are 92 countries in which naturally occurring nonhuman primates live, but just four—Brazil, the DRC, Indonesia and Madagascar—contain about 70 percent of the species. The species tables throughout this book indicate those countries where you might stand a good chance of seeing certain species, and some specific regions are listed below. When planning a primate-watching trip, try requesting information about where to find primates from the nearest embassy or tourist office of the country you intend to visit. Whether or not you find the information useful, every such query reminds the staff—and ultimately the country's government—that primates are an asset to be valued and protected. After your trip, write a letter of appreciation to the ambassador—by doing so you will contribute to a positive impact on conservation.

Africa

Bwindi Impenetrable Forest National Park, Uganda

See Bwindi Mountain Gorillas here; you may also see Black-and-white Colobus, red colobus, baboons and several guenon species.

Species total 12

Lopé National Park, Gabon

This is the site of the world's largest gatherings of nonhuman primates—up to 1,000 Mandrills at once. You can also see Black Colobus, Western Lowland Gorillas, Chimpanzees, mangabeys, Crowned Guenons and the endemic Sun-tailed Guenon, among others.

Species total 15

Nyungwe National Park, Rwanda

See huge troops of 300 or 400 Black-and-white Colobus, along with habituated Gray-cheeked Mangabeys, Blue Monkeys and Red-tailed Guenons; Owl-faced Monkeys are also present. You may have opportunities for Chimpanzee tracking.

Species total 13

Parc National de Mantadia-Andasibe, Madagascar

This national park is said to have "the biggest and most attractive lemurs habituated," including groups of Indri, Diademed Sifakas and Black-and-white Ruffed Lemurs, along with nine other lemur species.

Species total 12

Ranomafana National Park, Madagascar

This southeastern rainforest is home to a long-term study site, and many of the primate species living there are tolerant of visitors. You can view them quite close up, in the company of local guides. Mouse lemurs, the smallest primates in the world, will come out at night for pieces of banana.

Species total 13

Asia

Iwatayama Monkey Park, Arashiyama, Japan

Free-living Japanese Macaques are fed grain here for a long-term research program and as a public spectacle, on a hill overlooking Kyoto.

Species total 1

Kinabatangan Wildlife Sanctuary, Sabah, East Malaysia

From this sanctuary you can take riverboat rides to see Proboscis Monkeys, Bornean Orangutans and gibbons, Long-tailed and Pig-tailed Macaques; slow lorises and tarsiers are present but trickier to see.

Species total 10

Mentawai Islands, off west coast of Sumatra, Indonesia

Seven primates found nowhere else in the world live on these islands, which have been called "the Galapagos of Indonesia." They include Kloss's Gibbon, the Mentawai Macaque and the Mentawai Langur.

Species total 7

Polonnaruwa Archaeological Sanctuary and Horton Plains National Park, Sri Lanka

At Polonnaruwa you can look for gray langurs and the endemic Toque Macaque amid spectacular ruins. At Horton Plains you can track the endangered "bear monkey"—the shaggy montane form of the Purple-faced Langur—on a cold pre-breakfast walk or seek out slender lorises at night.

Species total 4

Tanjung Puting National Park, Indonesian Borneo (Kalimantan)

Here you can find Bornean Orangutans, Proboscis Monkeys, Long-tailed and Pig-tailed Macaques and, rarely seen, tarsiers.

Species total 8

The Americas

Mamiraua Ecological Station, Brazil

One day by boat from Manaus is the Uakari Floating Lodge, where you can watch monkeys, including the endangered White Bald Uakari, from canoes gliding through the flooded forest.

Species total 7

Manuel Antonio National Park, Costa Rica

In this small but biodiverse park you can see White-faced Capuchins, the Black-handed Spider Monkey and Red-backed Squirrel Monkey, and then go on to the dry tropical forests of coastal Guanacaste and the Nicoya Peninsula to hear and watch Venezuelan Red Howlers.

Species total 4

Manu National Park, Peru

In Manu you are likely to see eight primate species, including capuchins, the Common Squirrel Monkey, the Venezuelan Red Howler, two spider monkey species and the Emperor Tamarin. Other species, including Goeldi's Monkey and the Pygmy Marmoset, are present but trickier to spot. You can observe them at canopy level from two platforms 100 ft. (30 m) tall.

Species total 13

Poço das Antas Biological Reserve, Brazil

Golden Lion Tamarins bred in zoos have been reintroduced to this remnant of Atlantic coastal rainforest and now thrive here.

Species total 4

Further Resources

BOOKS

Caldecott, Julian, and Lera Miles (Eds). *World Atlas of Great Apes and their Conservation*. Berkeley: University of California Press, in association with UNEP-WCMC, 2005.

Campbell, Christina, Agustin Fuentes, Katherine Mackinnon, Melissa Panger and Simon Bearder (Eds). *Primates in Perspective*. New York: Oxford University Press USA, 2006.

Groves, Colin. *Primate Taxonomy*. Washington, D.C.: Smithsonian Institution Press, 2001.

Kavanagh, Michael. *A Complete Guide to Monkeys, Apes and Other Primates*. London: Jonathan Cape, 1983.

Kingdon, Jonathan. *The Kingdon Field Guide to African Mammals*. Princeton: Princeton University Press, 1997.

Macdonald, David (Ed). *Encyclopedia of Mammals*. New York: Facts on File, Inc., 2006.

Mills, Gus, and Lex Hes (Eds). *The Complete Book of Southern African Mammals*. Cape Town, South Africa: Struik, 1997.

Mittermeier, Russell et al. *Lemurs of Madagascar*. Arlington, Virginia: Conservation International, 2008.

Redmond, Ian. *Eyewitness Gorilla, Monkey and Ape*. New York: DK Publishing, Inc., 2000.

Rowe, Noel. *The Pictorial Guide to the Living Primates*. New York: Pogonias Press, 1996.

ONLINE INFORMATION AND MEDIA RESOURCES

Many websites can provide you with further information about primates and allow you to view images, ranging from spectacular photographs and videos taken by professional photographers to those taken by amateur enthusiasts of all ages. Some of the best of these websites are listed below, along with details of organizations you may wish to investigate that will allow you to become involved in helping to protect primates.

animaldiversity.ummz.umich.edu/
Animal Diversity Web (ADW) is an online encyclopedia of the natural history of animals at the University of Michigan. It contains thousands of detailed accounts of individual animal species as well as information about animal classification.

www.arkive.org/
ARKive gathers together films, photographs and audio recordings of the world's 16,000-plus species currently threatened with extinction into one centralized digital library. The database also provides profiles for many of these species.

www.bucknell.edu/msw3/
The third edition of Wilson and Reeder's book *Mammal Species of the World* is summarized in this online database of mammalian taxonomy. It provides you with the taxonomy, basic information and conservation status of all species of mammals.

www.cites.org
The Convention on International Trade in Endangered Species of Wild Fauna and Flora (CITES) has been signed by 172 countries. It is designed to control trade in wildlife and ensure that species are not threatened by trade, and this website keeps you up to date with CITES activities.

www.iucnredlist.org
This is the website of the IUCN (International Union for the Conservation of Nature and Natural Resources) Red List of Threatened and Endangered Species. It lists the endangered status of the world's species and subspecies as evaluated by the Species Survival Commission's Specialist Groups.

www.nationalgeographic.com/
The National Geographic Society (NGS) is a nonprofit educational organization that has for decades funded important research on primates in their natural habitat, including studies by Jane Goodall, Dian Fossey and Birute Galdikas on Chimpanzees, Mountain Gorillas and Bornean Orangutans, respectively. On this website you can find links to a huge resource of various NGS media such as magazines, articles, photographs and videos, including an online version of *National Geographic* magazine.

www.primate-sg.org/diversity.htm
The IUCN's Primate Specialist Group has its own site, which lists the 390 primate species and 649 subspecies recognized by its experts. This page tells you the regions where the 35 percent of primate species and subspecies considered to be threatened are located.

www.theprimata.com
This site contains useful fact sheets on most species of primates.

ORGANIZATIONS WORKING TO PROTECT PRIMATES

www.4apes.com/
This site provides links to about 80 NGOs (nongovernmental organizations) involved in ape conservation, working together as the Ape Alliance.

www.conservation.org
Conservation International is a nonprofit organization that supports the conservation of all species, focusing on those in biodiversity hotspots. It is known for its partnerships with local nongovernmental organizations and indigenous peoples, and to date has ensured that more than 300,000 square miles (800,000 sq. km) on land and at sea are protected.

www.ippl.org/
The International Primate Protection League is a global organization dedicated to fighting the trade in primates, caring for unwanted and abandoned primates and lobbying for better protection for primates in both wild and captive settings.

www.pasaprimates.org/
The Pan African Sanctuary Alliance links 18 primate sanctuaries across Africa, which care for and, where possible, rehabilitate back to the wild rescued and confiscated apes and monkeys.

www.unep.org/grasp
Details of the United Nations Great Apes Survival Project—a partnership of governments, UN bodies, NGOs, experts and the private sector—are given here, with maps, fact sheets and links to partners.

Glossary

alpha male/female The most dominant male or female in a group of animals.

anthropoid Describes animals that appear humanlike.

arboreal Describes creatures that live in, or spend much time in, trees; tree-dwelling.

bachelor group An all-male group, usually consisting of young males who have not yet acquired females.

belch vocalization A contact call made by gorillas when feeding in dense vegetation, which helps them keep in touch with other group members; it sounds like an extended human belch or like throat-clearing: "*harrrum-mwah.*"

brachiation Locomotion achieved by swinging from arm to arm beneath branches.

catarrhine primates Old World monkeys and apes, whose nostrils are close together and open downward.

cathemeral Describes animals that are active by both day and night.

cecotrophy The eating of feces when first passed in order to digest the food a second time, practiced by rabbits and some primates.

CITES The UN Convention on International Trade in Endangered Species of Wild Fauna and Flora, which regulates trade in animals, plants and their parts across borders and which bans the trade if it threatens the species' survival.

convergent evolution The independent evolution of a similar feature or features in two species that are not closely related; the opposite of *divergent evolution*.

Cro-Magnon human Early kind of *Homo sapiens* who lived in rock shelters beneath cliffs in France 40,000 years ago.

cryptic species Species that (to the human eye at least) look very similar to each other.

culture A body of behaviors or traditions that is transmitted by learning from one individual or generation to another.

divergent evolution The evolution of two or more species from a single species—a common ancestor. For example, differences may arise in reproductively isolated populations over many generations, perhaps in response to differing environments, causing them to diverge.

DNA–DNA hybridization A technique in molecular biology used to measure the closeness of the relationship, known as genetic distance, between two species.

dominance hierarchy A social structure in a group of animals in which each individual defers to those of higher status and displaces those of lower status; may be apparent when gathering food, for example, or taking opportunities for reproduction. Also known as the pecking order because it was first described in chickens.

endemic Describes an organism that is native to, and found only in, a particular place—a habitat, region, island or country.

estrus call A vocalization made only when a female is in estrus (receptive to mating).

fission–fusion society A social system in which individuals meet up in temprorary groups then separate and join others, so that they spend time with many different community members.

gallery forest Forest along the sides of rivers and streams in an otherwise unwooded landscape.

generalist A species that is able to live in different habitats and adapt to different conditions.

genus The first part of a species' scientific name, which is always capitalized; closely related species are in the same genus (pl. genera).

great ape The term generally applied to Chimpanzees, Bonobos, gorillas and orangutans, which are all now included in the human family Hominidae, which leads some to conclude that humans are great apes too.

infanticide The killing of an infant, which is often the instinctive act of a new male taking over a female or group of females.

ischial callosities Horny pads of skin on the bottoms of some primates, which protect them while sitting; also known as sitting pads.

IUCN The International Union for the Conservation of Nature and Natural Resources, formerly known as the World Conservation Union, members of which include governments, nongovernmental organizations and academics.

lesser ape Gibbons and Siamangs may be classified as lesser apes to distinguish them from great apes.

loud call A vocalization intended to carry a long way; often used to defend or announce the occupation of a territory.

mast fruiting The simultaneous production by many trees of huge quantities of fruit for no clear seasonal or climatological reason.

matriline Also known as the mother's line, the matriline is an individual's line of descent that has been traced back through mother to grandmother to great grandmother and so on. In some primate species, prominent traits are inherited via the matriline. In some primate social hierarchies, social standing is also passed on from mother to daughter.

Neanderthal An extinct early species of human (*Homo neanderthalensis*) that appeared in Europe some 130,000 years ago and disappeared 30,000 years ago.

olfactory communication The use of scent to communicate.

olfactory lobe The lower front part of the cerebral hemisphere of the brain, which processes the sense of smell.

opposable thumb The thumb is the first (inside) digit of primates' hands; in most primates, it opposes the other four digits, enabling the hand to grasp a branch or object.

piloerection The process by which hairs stand on end, often because an animal is excited, frightened or cold; in humans this manifests as "goose bumps."

platyrrhine primates Monkeys found only in the New World (South and Central America). These primates have widely set nostrils that open sideways.

polygynous Describes a male that has two or more mates.

prehensile Able to grasp; often used to describe toes or a tail that can curl around a support such as a branch.

primary forest Natural forest that has never been logged, either commercially or by villagers.

primate A member of the Primata Order of mammals, comprising all the animals described in this book.

primatology The branch of zoology that deals with primates.

quadrumanual climbing A climbing technique that involves the use of feet with opposing big toes as well as hands when climbing.

rainforest Forest that has an annual rainfall of 69–79 in. (1,750–2,000 mm).

sebaceous glands Small glands (often associated with sweat glands) in the skin that secrete an oily substance called sebum.

secondary forest Forest that grows back on land on which primary forest has been cleared, whether by human action or a natural event such as a landslide.

sexual dimorphism A difference in shape or size between males and females of a species.

species A group of organisms that can breed together to produce young of the same kind.

subfamily A category between family and *genus*. Subfamily names end in -inae.

superspecies Closely related, recently diverged species that have separate, non-overlapping ranges.

taxon A name applied to any organism or group of organisms, from subspecies to Phyla; taxonomy is the science of classifying living things.

vestigial Describes a structure in the body that appears to be the degenerate remains of an ancestral organ that no longer functions.

Index

Numbers in **bold** indicate pages where an in-depth treatment is given; numbers in *italics* indicate images.

Acknowledgments

This book has been very much a team effort, and I am grateful for the help and guidance of many friends and colleagues. Any mistakes or omissions, however, are down to me and I would be glad to receive any comments and corrections. Material that for reasons of space could not be included here will be available on a new website (www.primatefamilytree.com). I would particularly like to thank: Claudia Martin, for initiating the project; Kara Moses, who rescued me from impossible deadlines by researching and drafting text for lemurs, New World monkeys and some of the Old World monkeys chapter; my assistant Katy Jedamzik; and my long-suffering wife, Caroline, who entered most of the data in the tables and profile boxes. My editor Paul Docherty diligently checked the text for inconsistencies and clarity for the nonspecialist reader and remained calm in the face of my flexible approach to deadlines and frequent absences on conservation missions overseas; and Ivo Marloh created the impressive look of the book. I must also thank the outstanding design and editorial team of Hugh Schermuly, Cathy Meeus and Ben Hoare for doing such a great job in designing and editing the pages and creating the maps. Colleagues who have checked particular sections and/or offered photographs for selection include Hardi Baktiantoro, Richard Bergl, Mark Brazil, Caroline Harcourt, David Jay, Deepani Jayawardene, Ashley Leiman, Anna Nekaris, Rosek of Profauna, Noel Rowe, Felicia Ruperti, Anthony Rylands, Lee Kwok Shing, Gehan de Silva Wijeyeratne and Liz Williamson.

Ian Redmond Stroud, England, March 2008

PICTURE CREDITS

b = bottom **c** = center **t** = top **l** = left **r** = right

Front cover main image: Corbis; insets: **t** Corbis, **others** NPL
Spine NPL
Back cover **lt** Ian Redmond; **lb** © Michael Neugebauer; **rt** Ian Redmond; **rc** NPL; **rb** Ian Redmond
Back flap Ian Redmond

All photographic images inside this book were courtesy of Nature Picture Library, Bristol, U.K., except for the following:

Pages 7 © Michael Neugebauer; **11** Ian Redmond; **14bcl** Mark Brazil; **15br** Stefan Merker; **19b** Ian Redmond; **20** © 2005 Breuer T. Ndoundou-Hockemba, M., Fishlock V. taken from First Observation of Tool Use in Wild Gorillas Biol 3(11): e380/ Public Library of Science; **21** FLPA; **26** Science Photo Library; **26** Marshall Editons/Public Domain; **29tr** Science Photo Library; **30bl** FLPA; **35t** Scala, Florence; **35b** Ian Redmond; **36** Ian Redmond; **40** Lee Kwok Shing; **41b** inset & **41t** Ian Redmond; **42** Ardea; **45** Ian Redmond; **49** Corbis; **60** NHPA; **69** NHPA; **73** Stefan Merker; **84**, **94**, **95** & **97** Mark Brazil; **104** iStock; **118** & **119** Ian Redmond; **134t** Florian Möllers; **137** & **138** FLPA; **147** Ian Redmond; **148t** & **150** Corbis; **157** Ian Redmond; **160** FLPA; **161b** Ian Redmond; **166** FLPA; **168** Getty Images